What's Inside

A Year's Worth of 20-Minute Science!

Human Body

Chew and Swallow

Materials for each pair of students:
8½" x 11" sheet of paper
drinking straw
round balloon
6" piece of tape

Help students recognize the important role teeth and saliva play in getting food to the stomach with this simple experiment. Pair students. Then tell each duo that the straw represents an esophagus and the balloon represents a stomach. Have each twosome tape the balloon onto the straw as shown. Challenge each twosome to come up with a way to get the piece of paper into the balloon without removing the tape. To conclude, invite student pairs to explain their results and conclusions to the rest of the class. *(Students should discover that the paper needs to be torn into small pieces, which represents chewing. Students may also suggest that water, representing saliva, be used to transport the paper more easily to the balloon.)*

Super Saliva

Materials for each student:
small piece of a cracker, popcorn, bread, an apple, a carrot, and chocolate

Explain to students that saliva is excreted in the mouth and contains special enzymes to begin digestion. To illustrate this point, have each child put one food sample at a time in his mouth and hold it there for several minutes. Then ask students whether the food has become softer and started to dissolve or stayed the same. Have each child record on paper the changes (if any) he notices in each food sample. Challenge students to draw conclusions as to why some of the foods changed while others stayed the same. Tell students that saliva contains an enzyme called *amylase*. Explain that amylase breaks down starches and sugars, which is why the cracker, popcorn, bread, and chocolate should have started to dissolve in their mouths right away.

Life Science

4 20-Minute Science • ©The Mailbox® Books • TEC61365

Forms of Energy

What a Stretch!

Materials for each group of students:
copy of the top half of page 66
3 same-size rubber bands
ruler
yardstick

Help students understand potential and kinetic energy with rubber bands! In advance, make a masking tape starting line on the classroom floor. Then explain to students that kinetic energy is the energy of motion and potential energy is energy that is stored. Demonstrate to students how to stretch the rubber band as shown. Tell students that the stretched rubber band is an example of potential energy, and that releasing the rubber band is an example of kinetic energy. Before students begin the experiment, remind them to aim their rulers straight ahead and to avoid releasing a rubber band if a classmate is in its path. Direct students to complete the experiment on page 66. After the experiment, lead students in discussing how the stretch of the rubber band affects the distance it travels. *(As you stretch the rubber band, you increase its potential energy. This results in a greater kinetic energy, which makes the rubber band travel faster when released.)*

Swing Time

Materials for each pair of students:
copy of the bottom half of page 66
5 metal washers
16" length of string
unsharpened pencil
jumbo paper clip
masking tape

This idea not only demonstrates potential and kinetic energy, but also explores the science behind a pendulum! Demonstrate the steps on page 66, pointing out to students that the pencil, string, and paper clip make a pendulum. Then have each pair follow the steps on page 66 to complete the experiment. When students are finished, explain that the potential energy in the experiment is stored in the paper clip and washers. When the paper clip and washers are released, their potential energy converts into kinetic energy. Discuss the results of the experiment and lead students to understand that the number of swings stays basically constant regardless of the number of washers used. Explain that this happens because a pendulum will take the same amount of time to make every swing regardless of how heavy the weight at the end of it is.

Did You Know?
In a grandfather clock, a pendulum of 39 inches will swing back and forth 60 times in one minute, resulting in an accurate measure of time.

Physical Science

61

Weather & Climate

Blowing in the Wind

Materials for each pair of students:
copies of page 96 and page 97

Explain to students that wind speed is a major factor in the strength of a storm and that the Beaufort Scale is a tool used to categorize these speeds. Pair students and have each child look at a copy of page 96 as you discuss the Beaufort Scale. Next, have each pair cut out the game cards from page 97 and lay them out face down. Direct each duo to play a game similar to Concentration. In this game, a match is made when an observation card and a matching Beaufort Scale card are flipped. (Have students use page 96 to help them determine matches.) The child with more cards at the end of the game (or playing time) is the winner.

Measuring Air Pressure

Tell students that understanding changes in air pressure helps predict when a storm will occur. Explain that when air pressure is high, the weather is often clear; when air pressure is low, the weather is often stormy. Then group students and lead them through the following steps to make barometers.

Materials for each group of students:
plastic or glass jar
piece of cardboard
drinking straw
balloon
strong rubber band
tape
ruler

Steps for students:
1. Cut off the balloon nozzle. Gently stretch the larger opening across the mouth of the jar.
2. Secure the balloon in place with a rubber band.
3. Cut the straw at an angle to make a pointed end. Tape the uncut section of the straw to the flat balloon surface.
4. Hold the cardboard next to the jar. Where the straw points, mark the cardboard. Then, at the mark, draw a horizontal line across the cardboard.
5. Draw three lines above and three lines below the drawn line, each one-fourth inch apart. Label the top three lines "high" and the bottom three lines "low."
6. Place the barometer on a shelf or table near a wall. Tape the cardboard piece to the wall so that the pointed end of the straw aligns with the middle line.

Observation: Each day for a week (or longer if there hasn't been any rain in your area), have each group record the date, whether the air pressure is low or high, and the outside weather conditions. Lead students in a review of the relationship between air pressure and the type of weather it signals.

Earth & Space Science

86 20-Minute Science • ©The Mailbox® Books • TEC61365

Timesaving Formats!

- Material lists
- Step-by-step directions
- Quick-to-read descriptions
- Helpful illustrations

Chew and Swallow

Materials for each pair of students:
8½" x 11" sheet of paper
drinking straw
round balloon
6" piece of tape

Help students recognize the important role teeth and saliva play in getting food to the stomach with this simple experiment. Pair students. Then tell each duo that the straw represents an esophagus and the balloon represents a stomach. Have each twosome tape the balloon onto the straw as shown. Challenge each twosome to come up with a way to get the piece of paper into the balloon without removing the tape. To conclude, invite student pairs to explain their results and conclusions to the rest of the class. *(Students should discover that the paper needs to be torn into small pieces, which represents chewing. Students may also suggest that water, representing saliva, be used to transport the paper more easily to the balloon.)*

Super Saliva

Materials for each student:
small piece of a cracker, popcorn, bread, an apple, a carrot, and chocolate

Explain to students that saliva is excreted in the mouth and contains special enzymes to begin digestion. To illustrate this point, have each child put one food sample at a time in his mouth and hold it there for several minutes. Then ask students whether the food has become softer and started to dissolve or stayed the same. Have each child record on paper the changes (if any) he notices in each food sample. Challenge students to draw conclusions as to why some of the foods changed while others stayed the same. Tell students that saliva contains an enzyme called *amylase.* Explain that amylase breaks down starches and sugars, which is why the cracker, popcorn, bread, and chocolate should have started to dissolve in their mouths right away.

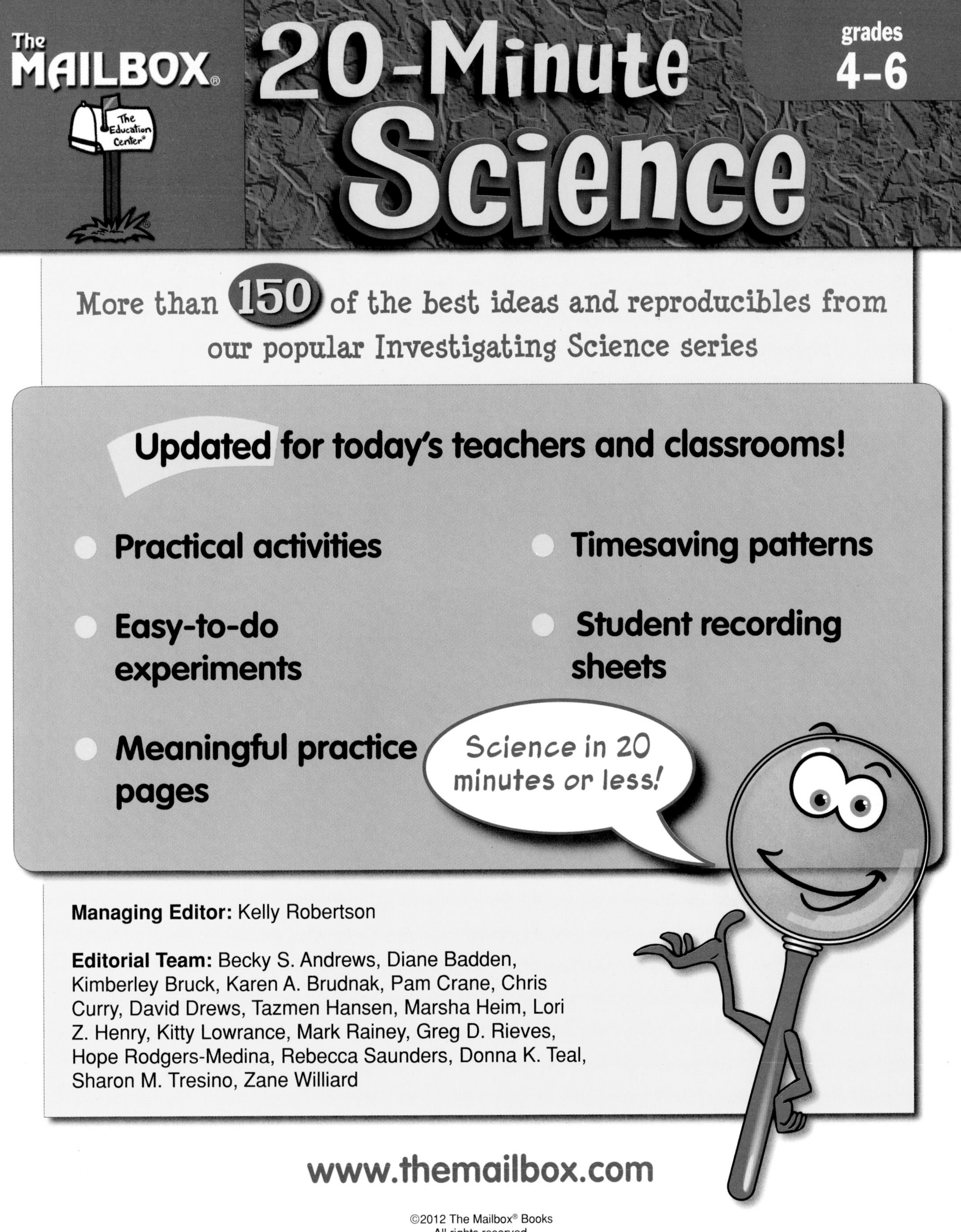

Managing Editor: Kelly Robertson

Editorial Team: Becky S. Andrews, Diane Badden, Kimberley Bruck, Karen A. Brudnak, Pam Crane, Chris Curry, David Drews, Tazmen Hansen, Marsha Heim, Lori Z. Henry, Kitty Lowrance, Mark Rainey, Greg D. Rieves, Hope Rodgers-Medina, Rebecca Saunders, Donna K. Teal, Sharon M. Tresino, Zane Williard

www.themailbox.com

ISBN 978-1-61276-220-3

Printed in the United States
10 9 8 7 6 5 4 3 2 1

HPS233481

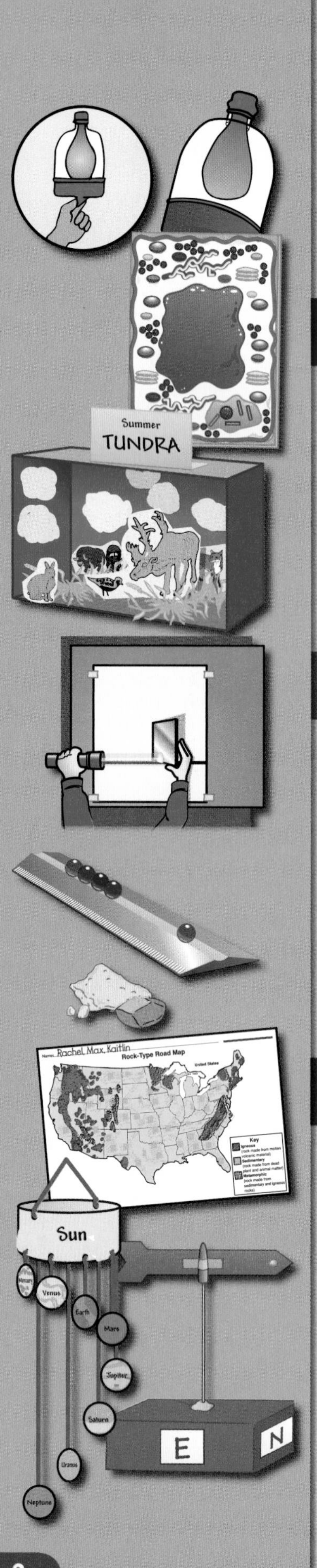

Table of Contents

Measuring the Digestive Tract

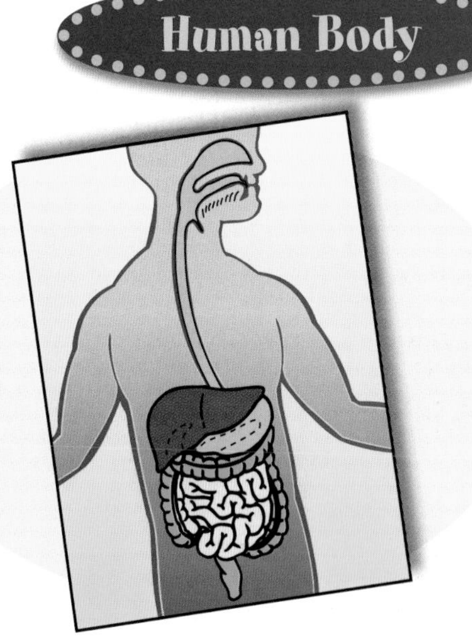

Help students visualize the length of the digestive tract with this activity. Follow the steps below to measure a student volunteer's digestive tract.

Materials:
ball of yarn
5 short lengths of different-colored yarn
yardstick or measuring tape

Steps:
1. Use the yarn from the ball to measure from the front (the corner of his lips) to the back (his jawbone) of the student's mouth. Tie a short piece of yarn at this point. Then, continuing with the yarn from the ball, complete each measurement in Steps 2–5.
2. Measure from the top of the neck to the stomach. Tie a short length of yarn at this point. This section of yarn represents the esophagus.
3. Have the student spread out the fingers on one of his hands. Measure the distance from the tip of his thumb to the tip of his little finger to represent the length of the stomach. Tie a short piece of yarn at this point.
4. Next, use the yarn to measure the student's height four times. Tie a piece of yarn here to represent the length of the small intestine.
5. Measure the student's height one time. Tie a piece of yarn here to represent the length of the large intestine.
6. Cut the yarn at the end of the large intestine. Spread out the entire length of the yarn and measure it to find the length of the student's digestive tract.

Did You Know?
The average length of the esophagus is ten inches.
The average length of the small intestine is about 19 feet, but it is coiled and looped tightly.
The average length of the large intestine is about five feet.
The average length of the entire digestive system is about 30 feet.

Cell Cartoons

Materials for each student:
copy of the blood cell information card from page 12
8½" x 11" sheet of white paper

Give your students an understanding of the vital roles of red and white blood cells with this activity. Instruct each child to fold his white paper into thirds, like a pamphlet, and then fold it in half to make six boxes. Have each student use his information card to write and illustrate a comic strip that shows the duties of either a white or a red blood cell.

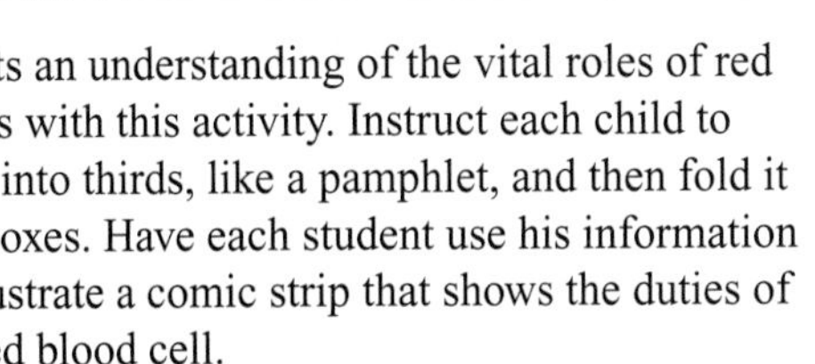

Bile Demonstration

Explain to students that bile, which is made by the liver and stored in the gallbladder, flows to the small intestine where it works to break down and absorb fat from food. Simulate the effect of bile for students with this small-group activity.

Materials for each group:
clear plastic cup
water
2 tbsp. oil
1 tbsp. of colored dishwashing liquid
spoon

Steps for students:
1. Put water in the cup, leaving one inch at the top.
2. Add the oil to the water. Observe the oil as it collects on top of the water. *(This simulates fat in the bloodstream.)*
3. Next, pour the dishwashing liquid in the water and oil mixture and stir. *(The dishwashing liquid simulates the bile that is secreted by the liver.)*
4. Observe what happens to the oil as the liquid begins to settle. *(The dishwashing liquid breaks the oil into smaller particles. This process is similar to the action of bile on fats. By breaking the fat down into smaller particles, the fat can be more readily absorbed by the body.)*

What's in Blood?

Share this demonstration to introduce students to an important part of the circulatory system.

Materials:
blood ingredients shown
large clear bowl
measuring cup
funnel
empty, clean quart beverage container

Steps:
1. Mix the blood ingredients in the bowl as you describe for students the main components of blood *(plasma, red blood cells, white blood cells,* and *platelets)*.
2. Use the funnel to pour the mixture into the quart container.
3. Display the container as you explain to students that a small child has one quart of blood in her body. As she grows into an adult, more red blood cells will be made in the bone marrow of some of her bones, and she will have five quarts of blood.

What's in Blood?

Plasma (54%)
- 2 c. water (representing the water in plasma)
- several drops yellow food coloring (protein)
- dash salt (salt)
- splash vegetable oil (fat)
- dash sugar (glucose)
- squirt honey (hormones)
- splash steak sauce (waste)

White Blood Cells and Platelets (1%)
- $\frac{1}{4}$ c. milk
- splash hot sauce (platelets)

Red Blood Cells (45%)
- $1\frac{3}{4}$ c. ketchup

Life Science

Pulse Meter

Here's a simple activity for teaching students about their pulse rates.

Materials:
class supply of page 13
toothpick
small ball of clay

Day 1: Help each student find his pulse by placing the index and middle fingers of one hand on the vessels of the opposite wrist. To make a pulse meter, direct each student to slightly flatten the ball of clay and then stick one end of the toothpick in the clay. Instruct him to rest his wrist on a desk or tabletop, place the pulse meter on the vessels of his wrist and observe what happens. *(The toothpick moves slightly.)* Next, have him remove the pulse meter, jump in place for a few seconds, return the pulse meter to his wrist, and observe what happens. *(The toothpick moves more visibly.)*

Day 2: Pair students. Have each student work with his partner to complete a copy of page 13.

Name ____________ Exploring pulse rates

The Beat Goes On

Your heart is a muscle about the size of your fist. It constantly pumps blood throughout your body. By counting the number of times blood surges through a vessel, you can tell your pulse rate, or how hard your heart is working. Work with a partner to complete the double-bar graph below.

To Calculate Pulse:
1. Count beats for 15 seconds.
2. Multiply this number by 4.

Example:
1. 25 beats in 15 seconds
2. 25 x 4 = 100 beats per minute

Look at the graph. Then answer each question below.

1. Which activities required the least work? sitting and standing
2. Which activities required the most work? running in place and jumping jacks
3. Predict what your pulse rate might be when you are asleep. My pulse rate might be 60 when I am sleeping.
4. Why is your pulse rate faster after running in place than after sitting still? Because my heart has to work harder when I'm exercising.
5. Predict how the graph results would change if each activity lasted longer. My heart would beat faster to pump more blood.
6. What is your favorite activity? Predict your pulse rate after completing this activity for an hour. Soccer is my favorite activity. My pulse rate might be 190 after playing.

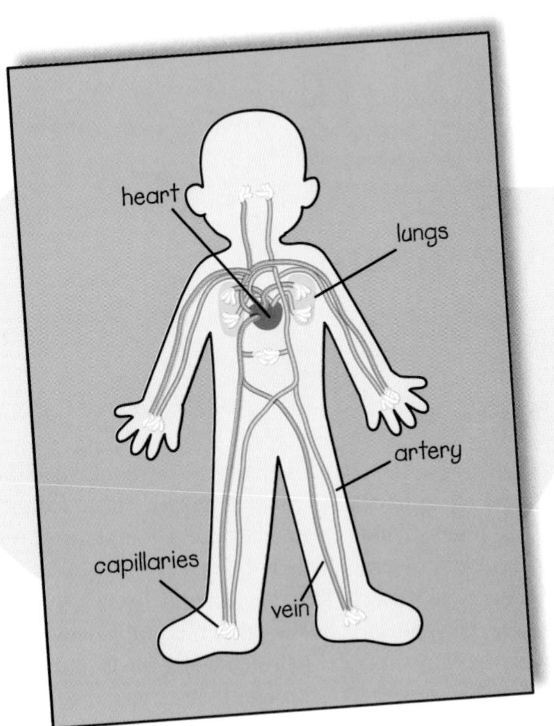

Circulatory System Model

Materials for each student:
enlarged copy of body pattern on page 12
red and yellow construction paper scraps
red, blue, and white yarn

Display an illustration of the circulatory system and point out to students the network of vessels (arteries, veins, and capillaries) that transport blood through the body. Instruct each child to cut out a heart and lungs from the construction paper scraps and glue them on the body outline as shown. Then have her glue the yarn (red for arteries, blue for veins, and white for capillaries) on the outline to represent the vessels. Direct each child to label her circulatory model as shown.

Breathing Teamwork

In advance, follow the steps shown to make a model to use to demonstrate how the diaphragm and lungs must work together for breathing to occur.

Materials:
clear plastic 20-ounce water bottle (a ribbed one works best)
two 12" balloons
rubber band
masking tape

Steps:

1. Cut off and recycle the bottom two-thirds of the bottle. Cover the cut edge of the remaining bottle with masking tape.
2. Push one of the balloons through the mouth of the bottle, keeping the balloon's mouth on the outside. Then stretch the mouth of the balloon over the mouth of the bottle to secure it.
3. Cut off the mouth of the other balloon. Then fit the rest of that balloon over the open bottom of the bottle. Secure the balloon with a rubber band.

To begin, explain to students that, because they aren't muscles, lungs can't move by themselves. Instead, they work with a large muscle at the bottom of the chest cavity called the *diaphragm.* Have a volunteer pull the drumlike bottom balloon outward with his thumb and forefinger (see illustration). Discuss with students what happens. *(Students should observe that, when the balloon bottom is pulled outward, more room is created inside the bottle. Increasing the room in the bottle decreases the air pressure. Air is then pulled into the balloon inside the bottle to make up for this change in air pressure. In the same way, when inhalation occurs, the diaphragm changes the volume of the chest cavity, causing the lungs to expand.)*

See the human body skill sheets on pages 14–16 and 18.

Reasons for Mucus

Materials:
2 pieces of clear plastic wrap
petroleum jelly
dust (collect from items such as bookshelves, computers, and desks)

Use this activity to demonstrate to students why we have mucus. Smear one piece of the plastic wrap with petroleum jelly. Explain to students that the petroleum jelly acts like mucus inside a person's nose. Blow some of the dust onto the plastic wrap with petroleum jelly and the rest onto the other piece of plastic wrap. Have the class observe that the dust and dirt is trapped in the petroleum jelly, but none has collected on the other piece of plastic wrap. Have students discuss why they think the petroleum jelly made a difference. Lead students to realize that, just as the petroleum jelly traps dust, the mucus inside our noses traps dirt and dust, preventing it from getting into our lungs.

Did You Know?
Mucus also traps germs and helps rid our bodies of them!

All Kinds of Bones

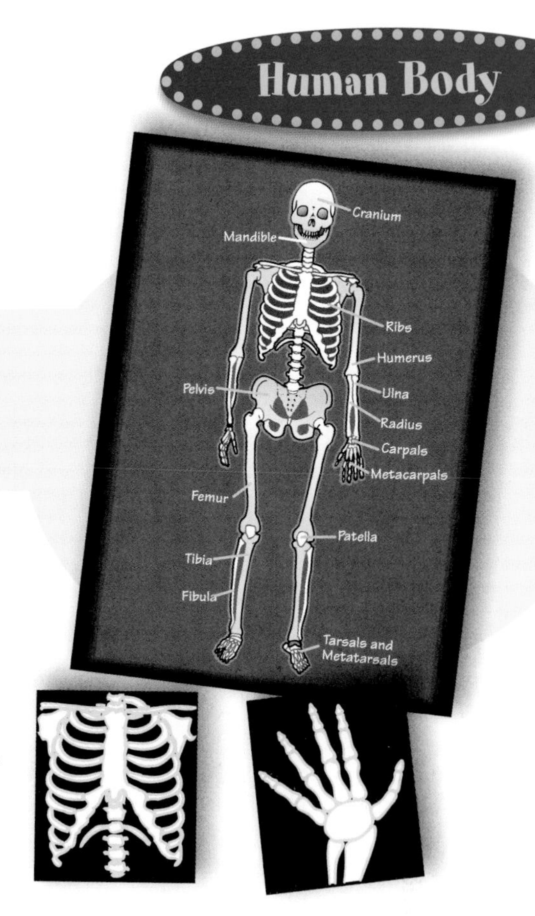

Materials:
pictures of X-rays (images printed from the internet, found in books, or actual discarded X-ray films)
skeletal reference (similar to the one shown)
access to reference materials

Day 1: Post the skeletal reference in an accessible area. Then divide students into small groups and give each group a picture of an X-ray. Direct students in the group to use the skeletal reference and other reference materials to identify and describe the functions of the bone(s) on their X-ray.

Day 2: Invite each group to display the picture of their X-ray as they name the bone(s) and describe their functions for the rest of the class.

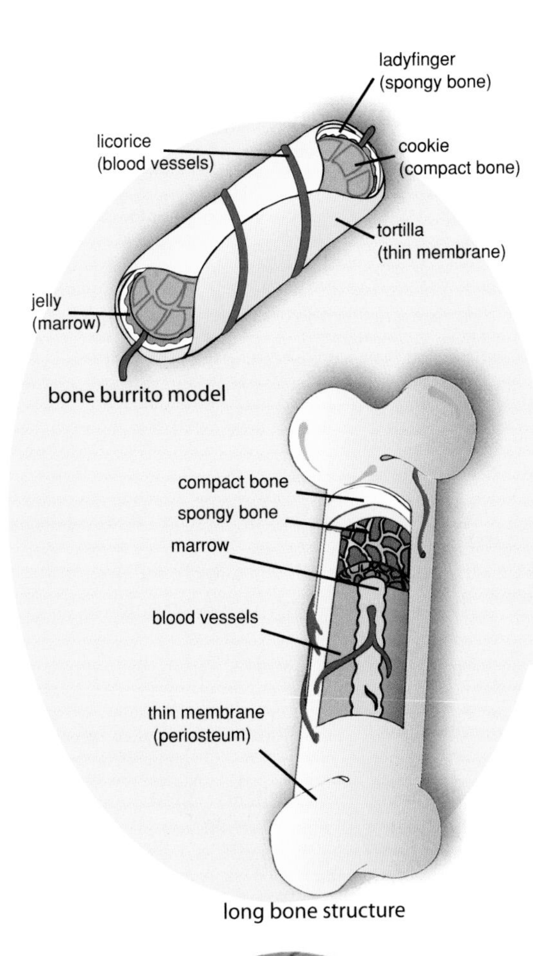

bone burrito model
long bone structure

Bone Burrito

Have your students make these edible bone models to help them visualize bone structure. Display a diagram of the layers of long bone, similar to the one shown, as you discuss with students the long bone structure. Then lead students through the steps shown to create edible bone models. When the models are complete, discuss the parts of the bone represented by each ingredient.

Materials for each student:
small soft tortilla
ladyfinger cookie
teaspoon of strawberry jam
plastic knife
oval-shaped sandwich cookie (such as a Vienna Finger cookie)
length of shoestring red licorice

Steps:
1. Open the ladyfinger and spread jam on it.
2. Lay one end of the licorice on the jam and close the ladyfinger.
3. Open the sandwich cookie and sandwich the ladyfinger between the halves.
4. Wrap the tortilla around the cookie layers and coil the remaining licorice around the tortilla.

A Neuron Network

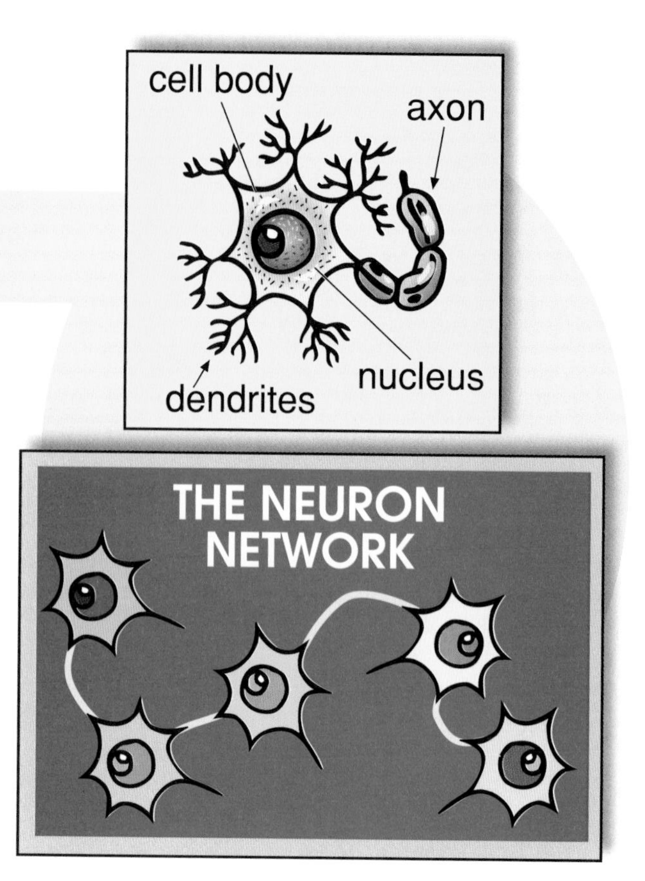

Materials for each student:
sheet of construction paper
length of yarn

Share with students that the average nerve cell, or neuron, is connected to at least 50,000 other neurons inside the human body. Explain that axons, strands that act like thin microscopic wires, pass nerve messages between neurons in tiny bursts of electricity. The junction between connected nerve cells is called a *synapse*. Show students a picture of a neuron, similar to the one shown. Then have each child refer to the picture to draw a neuron (complete with dendrites and a nucleus) on her paper. Direct her to cut out her neutron and tape the yarn to one dendrite to represent an axon. If desired, attach the axon of each child's neuron to a dendrite on another student's neuron on a display titled "The Neuron Network."

Did You Know?
There are three types of nerve cells: those that receive information from your senses, those that carry messages from your nervous system to different parts of your body, and those that send messages between cells.

Do You Remember?

Help students learn more about short-term memory with two simple experiments.

Materials:
class supply of page 17
10 different small classroom items in a paper bag

Steps:

1. Display the ten objects in a predetermined order.
2. Give students about a minute to study the order of the objects without touching them.
3. Direct students to look away or cover their eyes as you rearrange the order of the objects.
4. Have students list on their copies of page 17 what they remember of the original order of the objects.
5. Reveal the correct original order of the objects and have students check their answers.
6. Repeat Steps 2–5, using only seven objects.
7. Direct students to record their conclusions about the experiments where indicated. Then discuss the results of the experiments with students. *(It is easier to remember the order of seven objects than ten objects. This is because short-term memory can hold about seven items.)*

Name________________________

Digestive System Stars

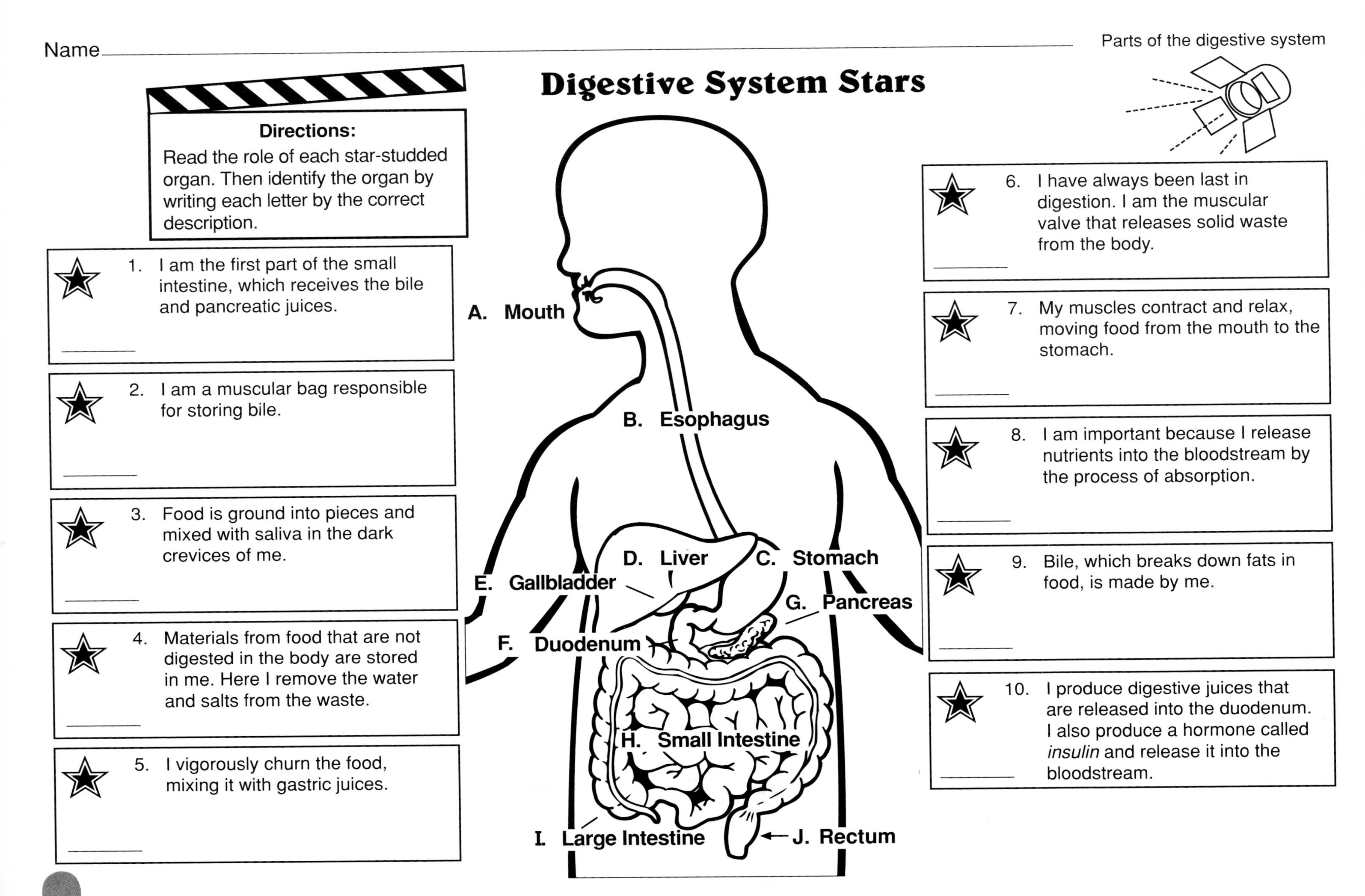

Directions:
Read the role of each star-studded organ. Then identify the organ by writing each letter by the correct description.

1. I am the first part of the small intestine, which receives the bile and pancreatic juices. ______
2. I am a muscular bag responsible for storing bile. ______
3. Food is ground into pieces and mixed with saliva in the dark crevices of me. ______
4. Materials from food that are not digested in the body are stored in me. Here I remove the water and salts from the waste. ______
5. I vigorously churn the food, mixing it with gastric juices. ______
6. I have always been last in digestion. I am the muscular valve that releases solid waste from the body. ______
7. My muscles contract and relax, moving food from the mouth to the stomach. ______
8. I am important because I release nutrients into the bloodstream by the process of absorption. ______
9. Bile, which breaks down fats in food, is made by me. ______
10. I produce digestive juices that are released into the duodenum. I also produce a hormone called *insulin* and release it into the bloodstream. ______

Blood Cell Information Card

Use with "Cell Cartoons" on page 5.

Red Blood Cells

There are billions of red blood cells in the body. They are shaped like tiny doughnuts but with dents instead of holes. Red blood cells pick up oxygen from the lungs, travel through the heart, and are pumped to the rest of the body through the arteries, capillaries, and veins. The red blood cells deliver oxygen to all the other cells in the body. Then they pick up carbon dioxide, a waste product, from the body and deliver it back to the lungs to breathe it out.

White Blood Cells

White blood cells come in many shapes and sizes. Some travel through the body to protect it. If this type of white blood cell recognizes a foreign substance, it will attach to it and then prepare to attack by multiplying. The multiplied white blood cells become plasma cells with Y-shaped antibodies. The antibodies help locate and destroy the foreign substance, making the body well.

TEC61365

Body Pattern

Use with "Circulatory System Model" on page 7.

Name ______________________

The Beat Goes On

Your heart is a muscle about the size of your fist. It constantly pumps blood throughout your body. By counting the number of times blood surges through a vessel, you can tell your pulse rate, or how hard your heart is working. Work with a partner to complete the double-bar graph below.

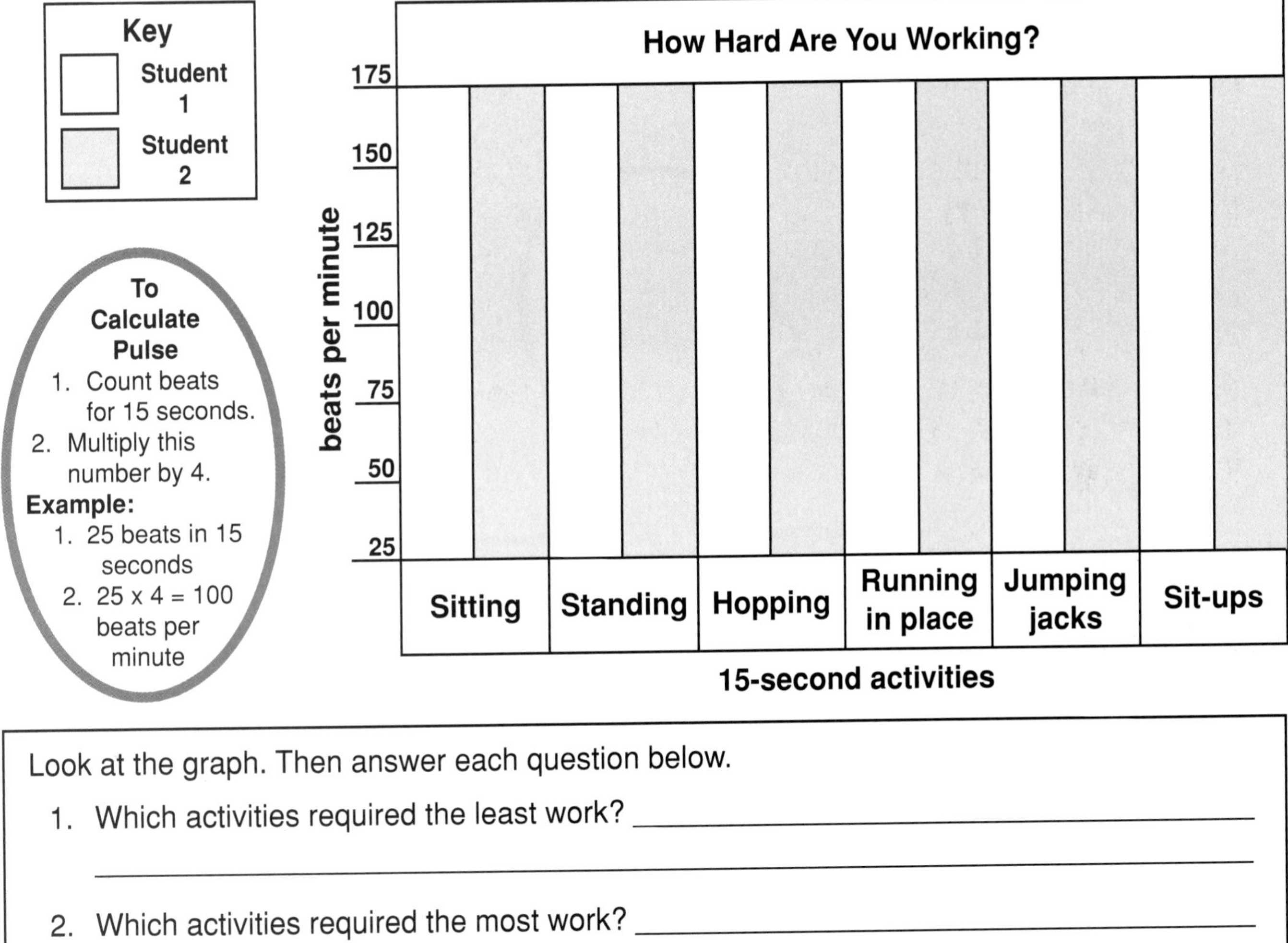

Look at the graph. Then answer each question below.

1. Which activities required the least work? ______________________
2. Which activities required the most work? ______________________
3. Predict what your pulse rate might be when you are asleep. ______________________
4. Why is your pulse rate faster after running in place than after sitting still? ______________________
5. Predict how the graph results would change if each activity lasted longer. ______________________
6. What is your favorite activity? Predict your pulse rate after completing this activity for an hour. ______________________

Note to the teacher: Use with "Pulse Meter" on page 7. Supply each pair with a stopwatch or a watch with a second hand.

Name ______________________ Identifying parts of the respiratory system

Breathe In, Breathe Out

Follow the directions below to create your own model of the respiratory system.

1. Match each organ to its description. Write the name of the organ in the correct blank and then cut out the strips.
2. Cut out the figure along the bold lines. Glue it onto a 9" x 12" sheet of construction paper.
3. Color and cut out each organ. Arrange each organ in the correct place on the figure and then glue the organs in place.
4. Glue each description strip beside its matching organ, drawing an arrow from the strip to its organ.

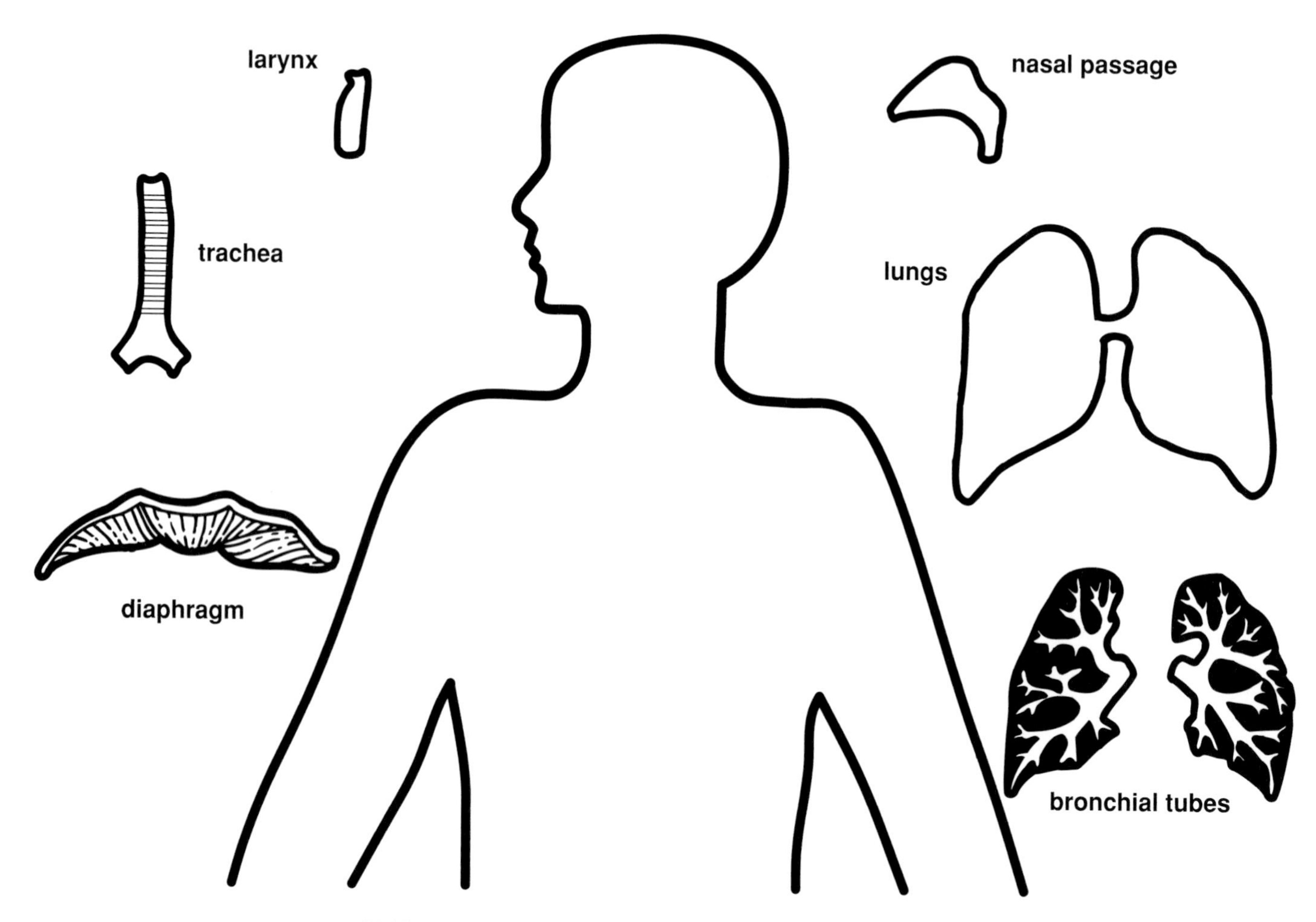

the place where oxygen-rich air enters the body ________	a muscle that contracts to allow the lungs to expand ________
passageway or soft tube from the mouth to the lungs ________	voice box at the top of the trachea that helps a person speak ________
the organs in which oxygen is exchanged with carbon dioxide ________	two tubes that lead from the trachea to the lungs ________

Note to the teacher: Provide each student with a copy of this page, one 9" x 12" sheet of construction paper, markers or crayons, glue, and scissors.

Name ____

Out of Breath!

Have you ever done an activity or exercise that left you totally breathless? Follow the directions for completing Parts I–III below to learn why this happens.

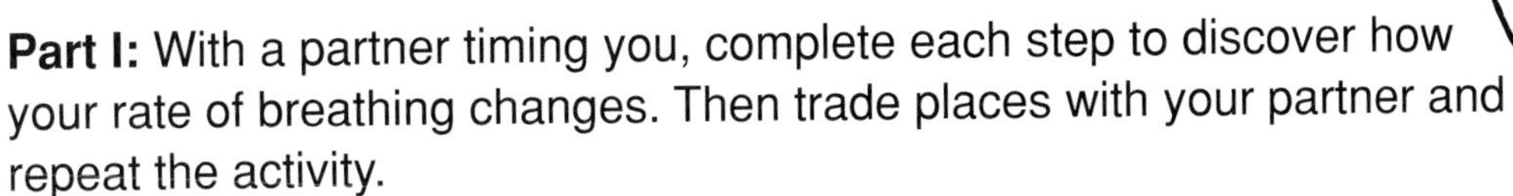

Part I: With a partner timing you, complete each step to discover how your rate of breathing changes. Then trade places with your partner and repeat the activity.

1. Sit in a chair quietly. With a partner timing you, breathe normally for one minute, counting the number of breaths you take. Record the amount. ________
2. Walk around the room at a normal pace for one minute. After one minute, continue walking for one more minute, this time counting the number of breaths you take. Record the amount. ________
3. Jog in place for one minute. After one minute, continue jogging for one more minute, this time counting the number of breaths you take. Record the amount. ________
4. Repeat Step 2. Record the amount. ________
5. Repeat Step 1. Record the amount. ________

Part II: On the line graph below, plot a point for the amount recorded in each step above. Then connect the points with lines.

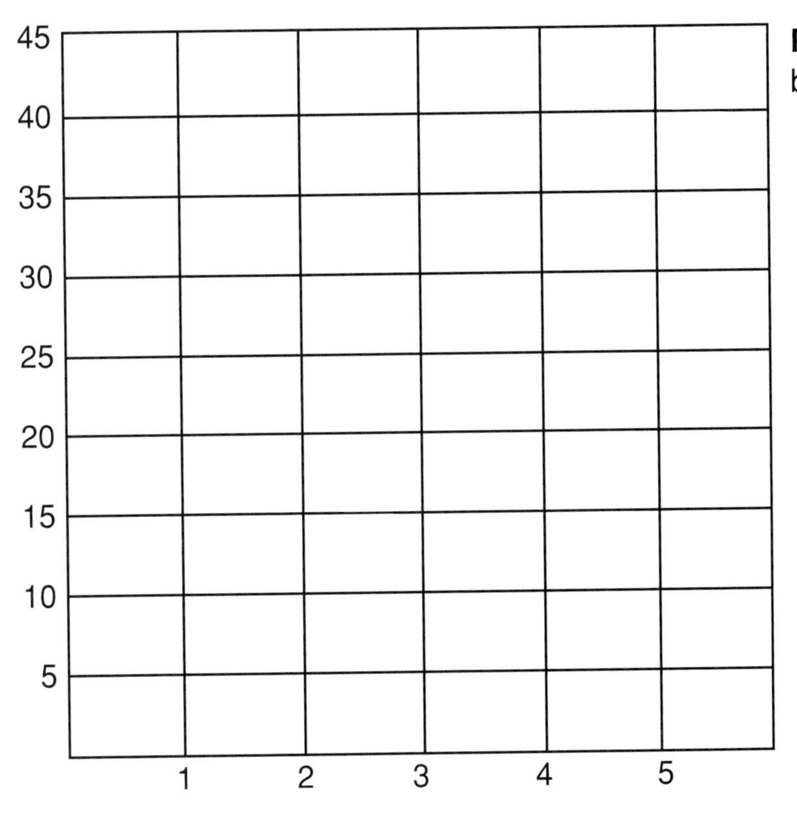

Part III: Answer the questions on the back of this sheet.

1. What did you observe about your breathing rate when you increased your activity? (Steps 1, 2, and 3)
2. What did you observe about your breathing rate when you decreased your activity? (Steps 4 and 5)
3. Why do you think a change in your breathing rate occurred when you increased or decreased your activity?

Bonus Box: Using a different color, plot on your line graph the amounts your partner recorded from the activity above. Compare the results.

Name ______________________________

All About Muscles!

Voluntary muscles—muscles you can control—are used for actions like writing, running, and talking. *Involuntary muscles*—muscles that work without your having to think about them—control actions like breathing and blinking.

Directions: Use the word bank to fill in each blank with the muscle action being described. Tell if the action is voluntary or involuntary by circling the appropriate letter. Then mark out that letter in the row at the bottom of the page. When you are finished, unscramble the remaining letters to discover the mystery muscle action that may be contagious.

Word Bank

coughing crying hiccuping adjusted your pupils smiling singing getting goose bumps sneezing chewing

1. You got nervous or ate too fast. A spasm in your diaphragm pulled air into your lungs through your voice box. You started ______________. It made quite a funny sound!

   voluntary—I involuntary—S

2. You received a ten-speed bicycle for your birthday. Two facial muscles pulled up the corners of your mouth and you started ______________.

   voluntary—C involuntary—A

3. You got out of the pool and felt cold. The muscles in your skin contracted, causing the hair to "stand on end." You started ______________!

   voluntary—G involuntary—Z

4. Your best friend gave you bubble gum before class. Powerful muscles in your cheeks and at the side of your head moved your jaw, and you started ______________.

   voluntary—R involuntary—Y

5. A tiny particle of food got caught in your windpipe. A muscle contraction sent a strong, sudden rush of air out from your lungs and your ______________ ejected the piece of food.

   voluntary—S involuntary—M

6. You moved from a dark hallway into the bright sunshine. Muscles known as sphincter muscles ______________ to the amount of light.

   voluntary—N involuntary—O

7. Pepper irritated some nerve endings. Muscles in your abdomen, chest, vocal cords, throat, and eyelids contracted. You began ______________.

   voluntary—N involuntary—L

8. Your vocal cords moved as air was forced through them. As the muscles got longer and shorter, they produced lower- and higher-pitched sounds, and you were ______________.

   voluntary—H involuntary—C

9. You laughed so long and hard that muscles around the *lacrimal glands* tightened. This caused salty fluid to be squeezed out, and you were ______________.

   voluntary—W involuntary—B

I	S	C	A	G	Z	R	Y	M	N	O	L	N	H	W	B

___ ___ ___ ___ ___ ___ ___

Name ______________________________

Memory Experiments

Experiment 1

Write the ten objects in their original order.

1. ____________ 2. ____________
3. ____________ 4. ____________
5. ____________ 6. ____________
7. ____________ 8. ____________
9. ____________ 10. ____________

Experiment 2

Write the seven objects in their original order.

1. ____________ 2. ____________
3. ____________ 4. ____________
5. ____________ 6. ____________
7. ____________

Conclusions

For which experiment did you list more items in the correct order? ________

Why do you think this is? ______________________________

Note to the teacher: Use with "Do You Remember?" on page 10.

Name ____

Sensory Nerves in Action

Study the diagram below. Then fill in each blank with a boldfaced word to complete the paragraph about how a nerve impulse travels from receptors in the skin to the brain.

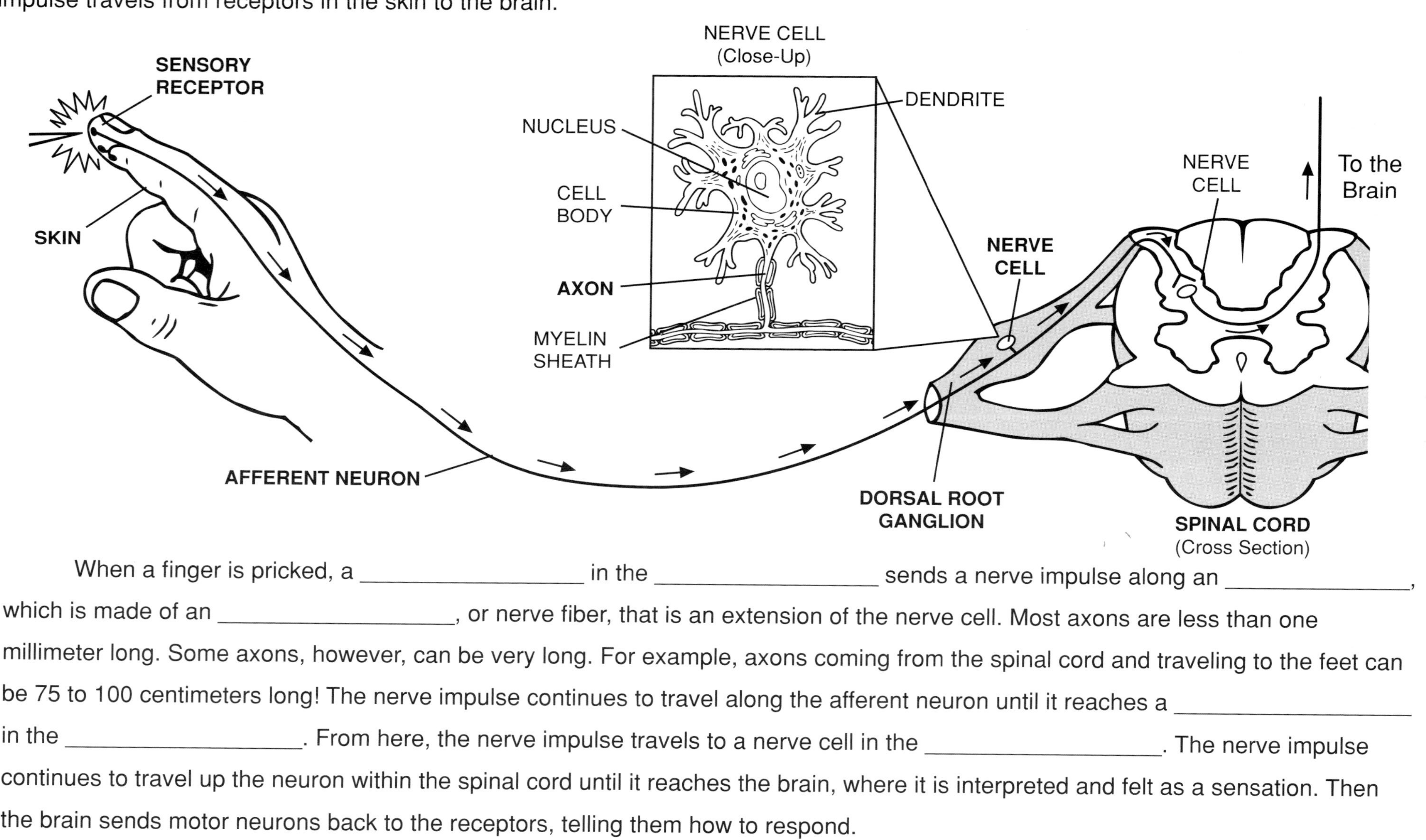

When a finger is pricked, a ________________ in the ________________ sends a nerve impulse along an ______________, which is made of an ________________, or nerve fiber, that is an extension of the nerve cell. Most axons are less than one millimeter long. Some axons, however, can be very long. For example, axons coming from the spinal cord and traveling to the feet can be 75 to 100 centimeters long! The nerve impulse continues to travel along the afferent neuron until it reaches a ________________ in the ________________. From here, the nerve impulse travels to a nerve cell in the ________________. The nerve impulse continues to travel up the neuron within the spinal cord until it reaches the brain, where it is interpreted and felt as a sensation. Then the brain sends motor neurons back to the receptors, telling them how to respond.

Plant Classification

Materials for each group:
copy of page 26
half sheet of posterboard
access to reference materials

Day 1: Explain to students that botanists consider plants to be one of the five kingdoms (overall groups) of living things. This kingdom is divided into ten divisions according to certain plant characteristics. Divide students into ten groups and assign each group a different division from the chart shown. Direct each group to use the reference materials to research its division and complete a copy of page 26.

Day 2: Instruct students in each group to use their notes from page 26 to list the information about their division on the posterboard and add illustrations. Display each project, as shown, to make a chart. Encourage students to refer to the chart throughout their study of the plant kingdom.

Conifers
- most have tall, straight trunks and narrow branches
- leaves in cooler climates are usually sharply tipped needles; flat narrow leaves; or tiny scalelike leaves
- most are evergreens
- keep their foliage all year long
- part of a group called gymnosperms
- seeds develop on woody scales of cones or inside fleshy cups
- male cones produce pollen
- Female cones stay on tree for many years

Plant Kingdom

Divisions

Conifers | Cycads | Flowering Plants | Horsetails | Ferns | Gingko | Club Mosses | Bryophytes | Whisk Ferns | Welwitschia, Ephedra & Gnetum

Fascinating Flowers

To help students identify the parts of a flower, lead them in making these 3-D blossoms.

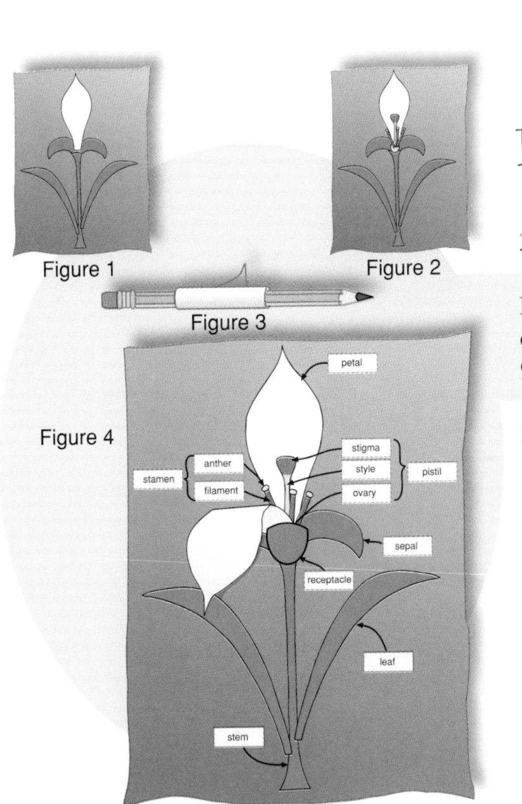

Materials for each pair of students:
copy of page 27
9" x 12" sheet of construction paper

Steps for students:
1. Color the flower parts on page 27. Cut out the flower parts and the word cards.
2. Glue the stem, sepal, one petal, and both leaves onto the construction paper (Figure 1).
3. Glue the stamens and the pistil on the petal (Figure 2).
4. Roll the other petal around a pencil (Figure 3).
5. Glue the bottom edge of the rolled petal to the top of the stem so the rolled side faces out. Then glue the receptacle over this petal's base (Figure 4).
6. Label each flower part with the corresponding word card.

Unusual Plants

Invite students to research some plants that have out-of-the-ordinary characteristics.

Materials for each pair of students:
access to reference materials
12" x 18" sheet of construction paper

Day 1: Remind students that most plants make their food during photosynthesis and normally get plenty of water and nutrients from soil. Then explain there are some plants that have evolved other methods of surviving. These plants are

- parasitic: attach themselves onto other plants called hosts and steal the hosts' nutrients and mineral supplies
- epiphytic: live on hosts—usually branches or stems—only to be closer to the sunlight
- carnivorous: get nutrients by capturing small animals and insects in their leaves

Next, pair students and assign each duo a plant from the list shown. Have each twosome begin researching its assigned plant.

Parasitic plants:	Carnivorous plants:	Epiphytic plants:
dodder plant	sundew plant	bromeliad
giant rafflesia	monkey cup pitcher	epiphytic orchid
ghost orchid	marsh pitcher	strangler fig
	yellow trumpet pitcher	epiphylls
	Venus flytrap	staghorn fern
	butterwort	
	bladderwort	

Day 2: Have each pair finish researching its assigned plant. Then direct the duo to make a simple wanted poster on a sheet of construction paper. Instruct the twosome to include what the plant looks like, how it survives, where it lives, and a picture of the plant. Bind the completed posters into a class book titled "Wanted: Strange and Unusual Plants!"

Plant Cell Structure

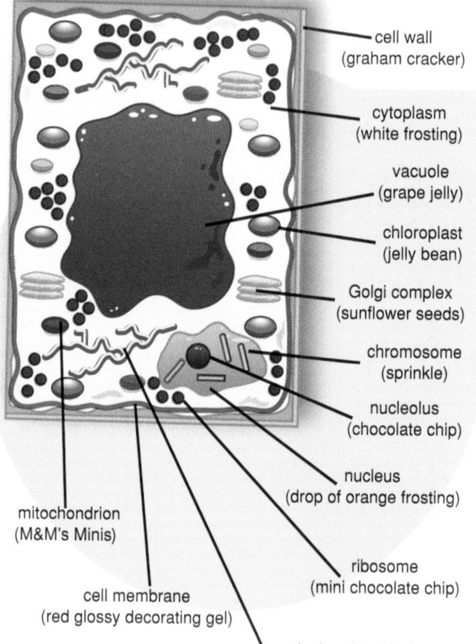

Materials for each student:

copy of page 28	white spreadable frosting
access to reference materials	various decorative frostings
graham cracker	assortment of decorative candies

Explain to students that cells are found in all living things. A plant cell's parts perform specific jobs, are enclosed by a cell wall that makes them rigid, and contain bright green organelles called *chloroplasts.* Review the plant parts on page 28 with students. Then have each child use reference materials to complete his copy of the page.

Optional follow-up: Place spreadable white frosting, graham crackers, and decorative frostings and candies at a center. Pair students and have each duo visit the center and use a completed reproducible to create a model of a plant cell similar to the one shown. When their model is complete, invite each twosome to identify each part of their plant cell and explain why they chose to represent it with the chosen food item. After students share their models, invite them to eat their creations.

Edible Plant Parts

Materials:
class supply of page 29
variety of edible plant parts, thoroughly washed (see list)
resealable plastic bag for each small group
paper towel for each small group

Day 1: Tell students that they eat many different plant parts—roots, stems, leaves, seeds, fruits, and flowers—every day. Then form small groups and give each group a resealable plastic bag containing one item to represent each plant part and a paper towel. Instruct the group to use its food items to form a "plant" on the paper towel as shown. After each group explains its model, invite them to munch on their creations.

Day 2: Review with students different foods that are plant parts. Then have each child complete a copy of page 29.

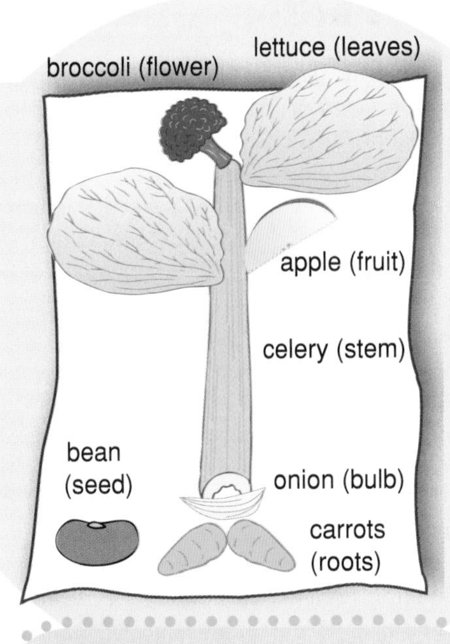

Edible parts of plants:
Roots: carrots, radishes, sweet potatoes
Bulbs: onions, garlic
Stems: asparagus, green onions, celery
Leaves: cabbage, lettuce, spinach
Flowers: broccoli, cauliflower
Fruits: apples, pears, peaches, cucumbers, pumpkins, tomatoes, peppers
Seeds: peas, beans, corn, peanuts

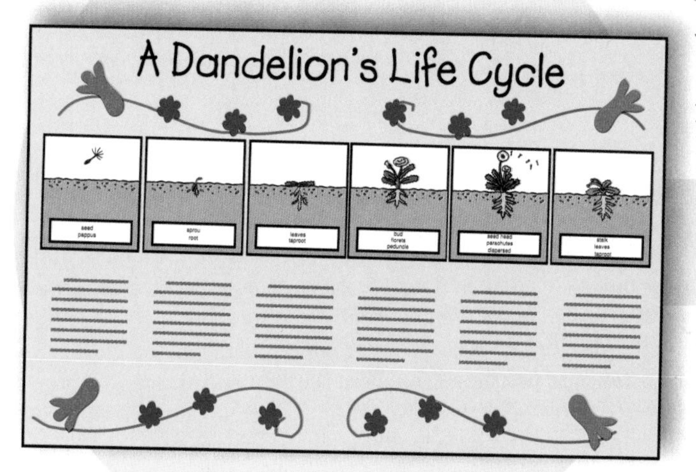

Dandelion Life Cycle

Discuss the life cycle of a dandelion with students. Then have students follow the steps below to review the dandelion life cycle.

Materials:
copy of page 30
12" x 18" sheet of construction paper

Steps for students:
1. Color and cut out the life cycle cards on page 30.
2. Put the cards in order and glue them on the paper.
3. Below each card write a description for each stage of the life cycle shown.
4. Add a title and desired decorations to your project.

Focus on Photosynthesis

Challenge students to describe how they could provide food for themselves using only four ingredients: sunlight, water, minerals from soil, and air. Then say that those ingredients are what plants use to make their own food. Explain that chlorophyll in a plant's leaves traps sunlight. Together with carbon dioxide from the air and minerals from the water that the roots take in, a plant makes food and oxygen. Next, pair students and guide them through the steps below to play a game to help them remember the four key elements of photosynthesis.

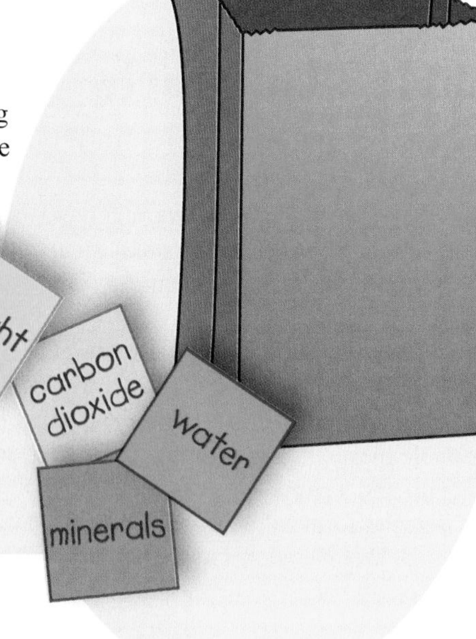

Materials for each pair of students:
paper lunch bag
4 small construction paper squares in each of the following colors and labeled as indicated: yellow, "sunlight," brown, "minerals," gray, "carbon dioxide," blue, "water"

Steps for students:

1. Place the cards in the bag.
2. Without looking, each player draws four cards from the bag.
3. If a player has a complete set of elements (sunlight, minerals, carbon dioxide, and water), she is the winner. If not, she returns one or more cards to the bag and picks a new card(s).
4. Play continues until one player has a complete set of elements.

Did You Know?
Plants breathe through their leaves and roots.

Enlightening Leaves

Materials:
leafy green plant, such as an African violet
construction paper trimmed to fit over the top and bottom of two of the plant's leaves
paper clips

Use this experiment to demonstrate to students that plants must have light to make food. Use paper clips to attach the construction paper to the tops and bottoms of two leaves. Then set the plant in direct sunlight. Ask each child to predict what he thinks will happen. After about a week, remove the papers from the leaves and ask students to compare the covered and uncovered leaves. *(The covered leaves should have yellowed and may have started to shrivel, while the uncovered leaves should have remained green.)* Guide students to realize that this is because the uncovered parts received more sunlight and absorbed more energy than the covered parts. With the leaves still uncovered, place the plant in direct sunlight again. Ask students what they think will happen. After another week or so, encourage students to observe the leaves. *(The leaves that had been covered should become green again.)* Lead students to conclude that chlorophyll works with energy from the sun to help carbon dioxide and water combine to make a plant's food. If a plant does not have food, it will eventually die.

Phenomenal Photosynthesis

Materials for each student:
copy of page 31
red, blue, and orange markers

Help students visualize the chemical change that takes place during photosynthesis with this colorful model. To begin, tell students that all substances are made of molecules and that molecules are made up of even smaller particles called atoms that can combine to form new substances. Direct each student's attention to the equation on page 31. Tell students that it is the chemical equation for photosynthesis (the food-making process of plants). Explain that each part of the equation represents different molecules: on the left, six molecules of carbon dioxide (6 CO_2) and six molecules of water (6 H_2O); on the right, one molecule of sugar ($C_6H_{12}O_6$) and six molecules of oxygen (6 O_2). Point out to students that, in photosynthesis, the molecules of carbon, oxygen, and hydrogen are not used up but combine to form new substances, including water, carbon dioxide, and sugar (glucose). Then have each student complete his copy of page 31.

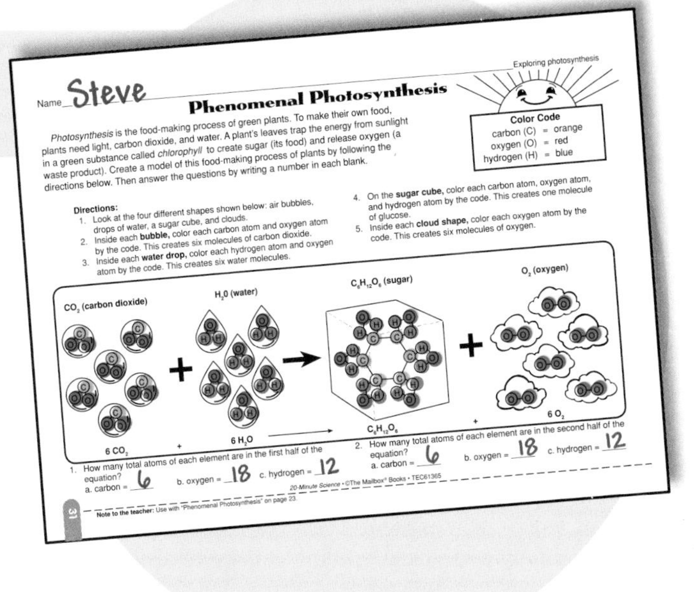
Exploring photosynthesis

Name Steve

Phenomenal Photosynthesis

Photosynthesis is the food-making process of green plants. To make their own food, plants need light, carbon dioxide, and water. A plant's leaves trap the energy from sunlight in a green substance called *chlorophyll* to create sugar (its food) and release oxygen (a waste product). Create a model of this food-making process of plants by following the directions below. Then answer the questions by writing a number in each blank.

Color Code
carbon (C) = orange
oxygen (O) = red
hydrogen (H) = blue

Directions:
1. Look at the four different shapes shown below: air bubbles, drops of water, a sugar cube, and clouds.
2. Inside each **bubble,** color each carbon atom and oxygen atom by the code. This creates six molecules of carbon dioxide.
3. Inside each **water drop,** color each hydrogen atom and oxygen atom by the code. This creates six water molecules.
4. On the **sugar cube,** color each carbon atom, oxygen atom, and hydrogen atom by the code. This creates one molecule of glucose.
5. Inside each **cloud shape,** color each oxygen atom by the code. This creates six molecules of oxygen.

1. How many total atoms of each element are in the first half of the equation?
   a. carbon = 6 b. oxygen = 18 c. hydrogen = 12
2. How many total atoms of each element are in the second half of the equation?
   a. carbon = 6 b. oxygen = 18 c. hydrogen = 12

20-Minute Science • ©The Mailbox® Books • TEC61365

Note to the teacher: Use with "Phenomenal Photosynthesis" on page 23.

Effects of the Sun

Materials for each student:
white paper plate
lined paper

Day 1: Discuss with students whether they think getting too much sun can be bad for living things. Brainstorm with students a list of positive and negative effects of the sun, listing students' responses on chart paper in two different columns. Then have each child decorate the rim of her paper plate on both sides so it resembles the sun.

Day 2: Direct each child to review the lists of the positive and negative effects of the sun. Then have her cut two circles from lined paper, both sized to fit the center of her plate. Have each student refer to the lists on the board as she writes one paragraph about the sun's positive effects and another about the negative effects. Have each child glue one paragraph to each side of her paper plate.

Effects of the Sun

Positive effects	Negative effects
provides light	can cause skin cancer
provides warmth	can cause heat stroke
provides solar energy	can damage the eyes if looked at directly
helps green plants grow and make food	can contribute to drought
helps water evaporate	can damage outer surfaces of some materials
helps grow plants for food	can burn the skin

Life Science

Moving On Up!

Use this demonstration to show students how water travels up from a plant's roots to the stems and leaves.

Materials:
class supply of page 32
5 oz. cup filled with water
paper towel
25 pennies
access to reference materials

Day 1: Show students the cup of water and have them estimate how many pennies can be added to the cup before the water spills over the top edge. Carefully add one penny at a time to the cup and have students observe what happens. *(The water forms a bulge that rises above the cup's rim.)* Explain to students that the bulge represents water's cohesiveness, or ability to stick together. Tell students that the cohesiveness of water molecules is an important factor in transpiration, the movement of water through a plant. Point out that scientists believe that as plants evaporate extra water through their stomata (tiny holes in their leaves), the entire column of water in a plant's stem is pulled upward through its xylem tubes.

Day 2: Review the process of transpiration with students. Then have each child complete a copy of page 32, using reference materials as needed.

Sweating Plants

Materials:
4 different types of same-size plants (smaller than a gallon-size resealable plastic bag)
4 gallon-size resealable plastic bags

To begin, ask students if they think plants sweat. Then explain that during transpiration, plants allow unused water to escape through the stomata on their leaves. Tell students that they will be observing how much water can be given off by common houseplants. Water each of the plants evenly; then place each in a bag and seal the bag. Place the bags in a sunny spot. After several hours, have students observe the water droplets that have condensed inside the bags and discuss whether some plants "sweat" more than others.

Did You Know?
On a hot day, a corn plant can release about a gallon of water.

Soil Sources

Materials for each group:
12" x 18" sheet of construction paper
9" x 12" sheet of construction paper
access to reference materials

Day 1: Tell students that soil is constantly being destroyed and reformed. Have students brainstorm sources of new soil *(sun, air, water, and other environmental forces that erode rock)* as you list their responses on chart paper.

Day 2: Review the list of soil sources with students. Then divide students into small groups. Have each group research several sources of soil to find out how those sources produce new soil.

Day 3: Have students use their research and the steps below to create a lift-the-flap project.

Steps for students:
1. From the smaller sheet of paper, cut four-inch squares (one for each soil source found). Label each square with a different soil source.
2. Arrange the squares on the larger sheet of paper; then label the larger sheet "Sources of Soil."
3. Glue the top edge of each square to the larger sheet of paper to form flaps.
4. On the paper under each flap, write a description or draw a picture to show how that source produces soil.

Delicious "Dirt"
1 large package Oreo® cookies, crushed
1 large package chocolate instant pudding
Gummy Worm candies®
small clear plastic cups
plastic spoons

Mix pudding according to package directions. In each cup, alternate layers of pudding and crushed cookies to make a total of four layers. Add a candy worm and serve.

Did You Know?
It can take 500 years for nature to create one inch of topsoil.

Hungry for Air

Materials:
2 plants (approximately the same size)
glass jar (large enough to cover one plant)
duct tape
ruler

This simple test helps students discover if a plant will starve if it is not exposed to air. To begin, measure and water the plants. Then place them in a sunny area, but not in direct sunlight. Turn the jar upside down over one plant and use duct tape to seal the jar to the surface it is sitting on.

Observation: Have students observe the plants for two weeks, and then measure the plants again. *(The plant under the jar should show no growth; the other plant should grow slightly.)*

Follow-up: Lead students to understand that green plants require carbon dioxide to carry on photosynthesis (their food-making process). If carbon dioxide is not present, the plant stops making food. Have students infer what would happen to the plant if the jar were not removed. *(The plant would eventually starve.)*

See the plant skill sheets on pages 33–36.

Name ______________________ Classifying plants

Plant Classification

Use reference materials to complete each section.

____

Plant Division

Note to the teacher: Use with "Plant Classification" on page 19.

Flower Part Patterns and Word Cards

Use with "Fascinating Flowers" on page 19.

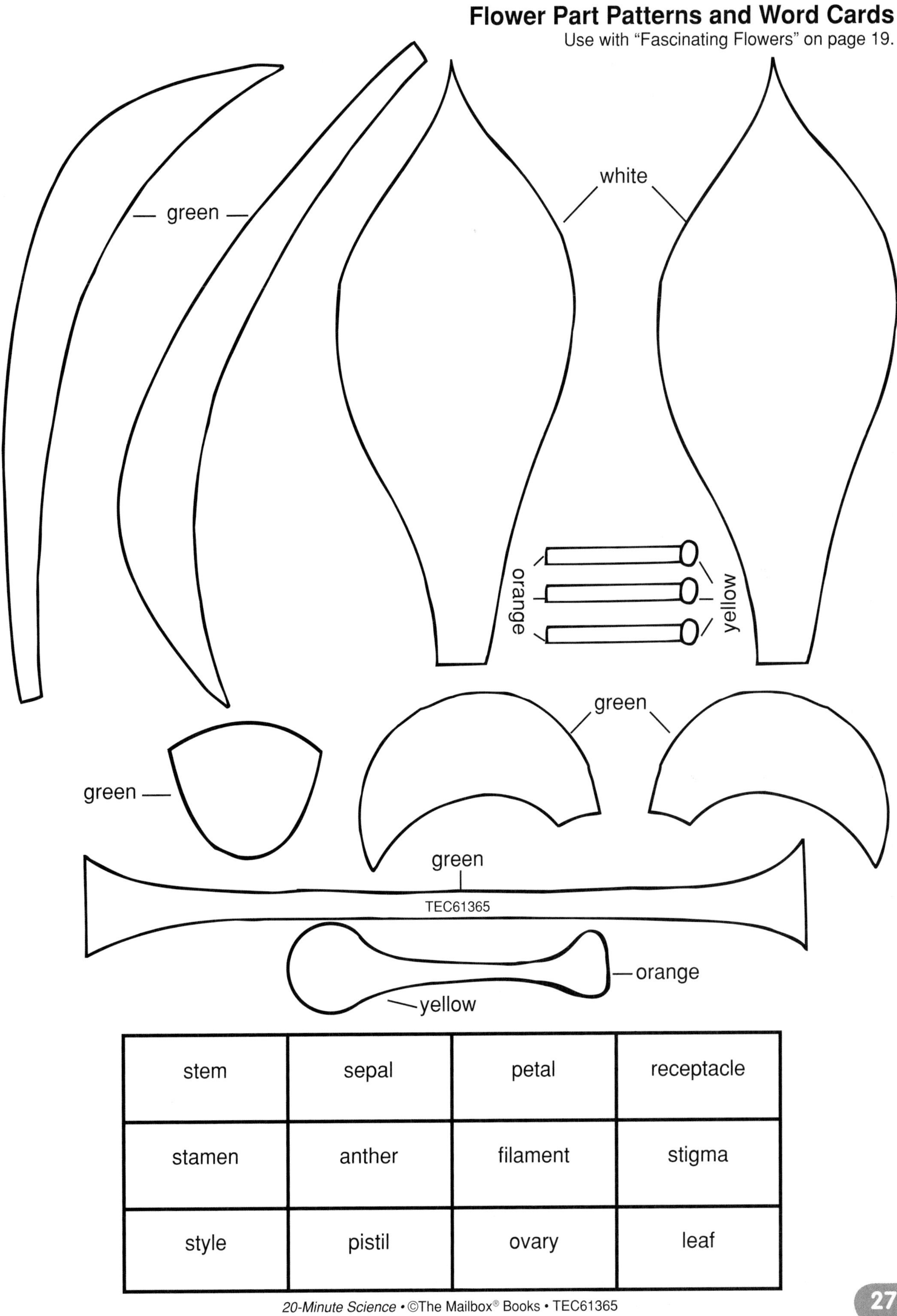

stem	sepal	petal	receptacle
stamen	anther	filament	stigma
style	pistil	ovary	leaf

Name ____________________

Cell Structure

Look at the diagram. Read each definition; then write the word that matches the definition.

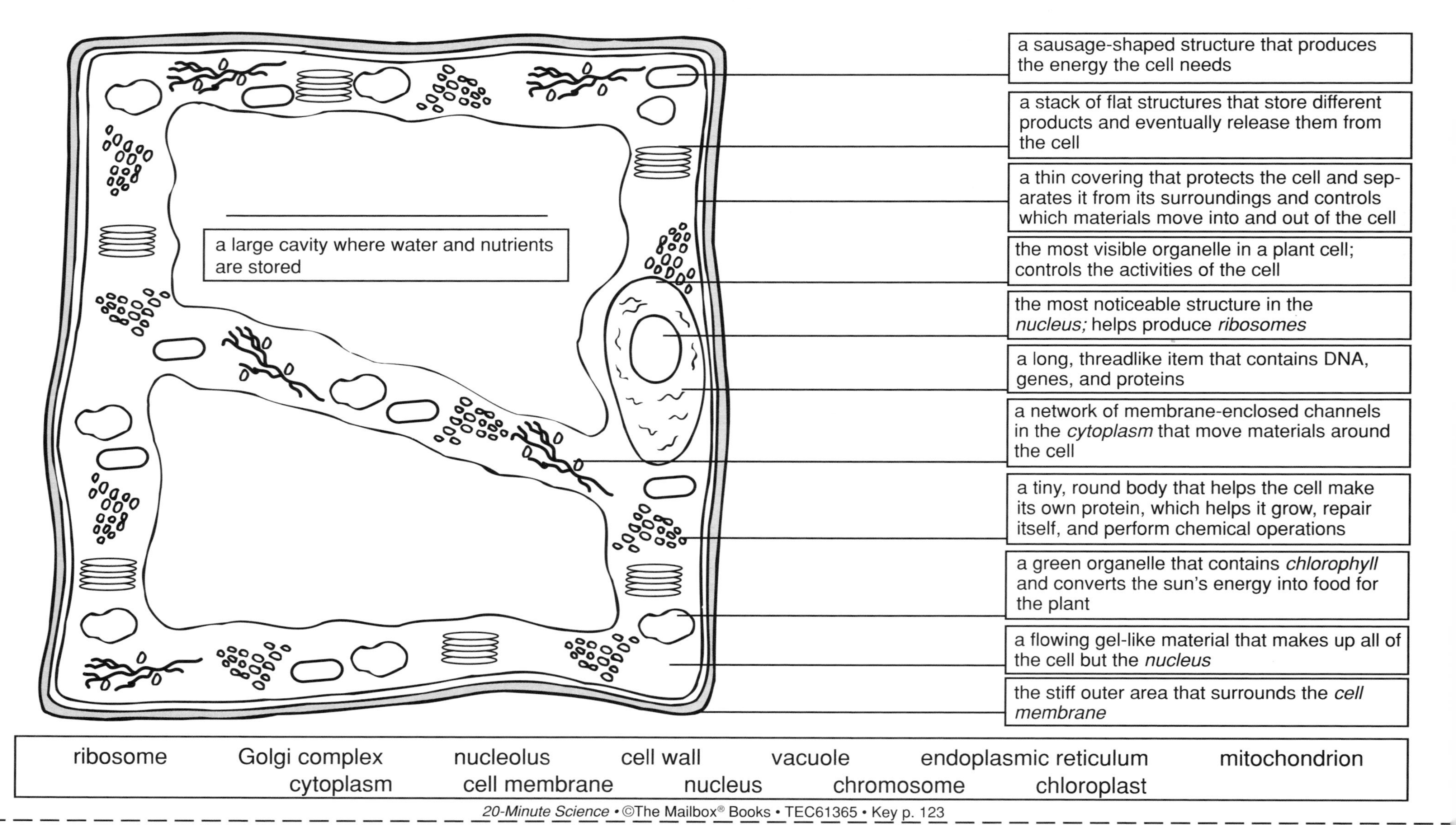

ribosome Golgi complex nucleolus cell wall vacuole endoplasmic reticulum mitochondrion
cytoplasm cell membrane nucleus chromosome chloroplast

Note to the teacher: Use with "Plant Cell Structure" on page 20.

Name ______________________________

Organizing Edibles

Color and cut out the picture cards below. Glue each one by its label on the chart.

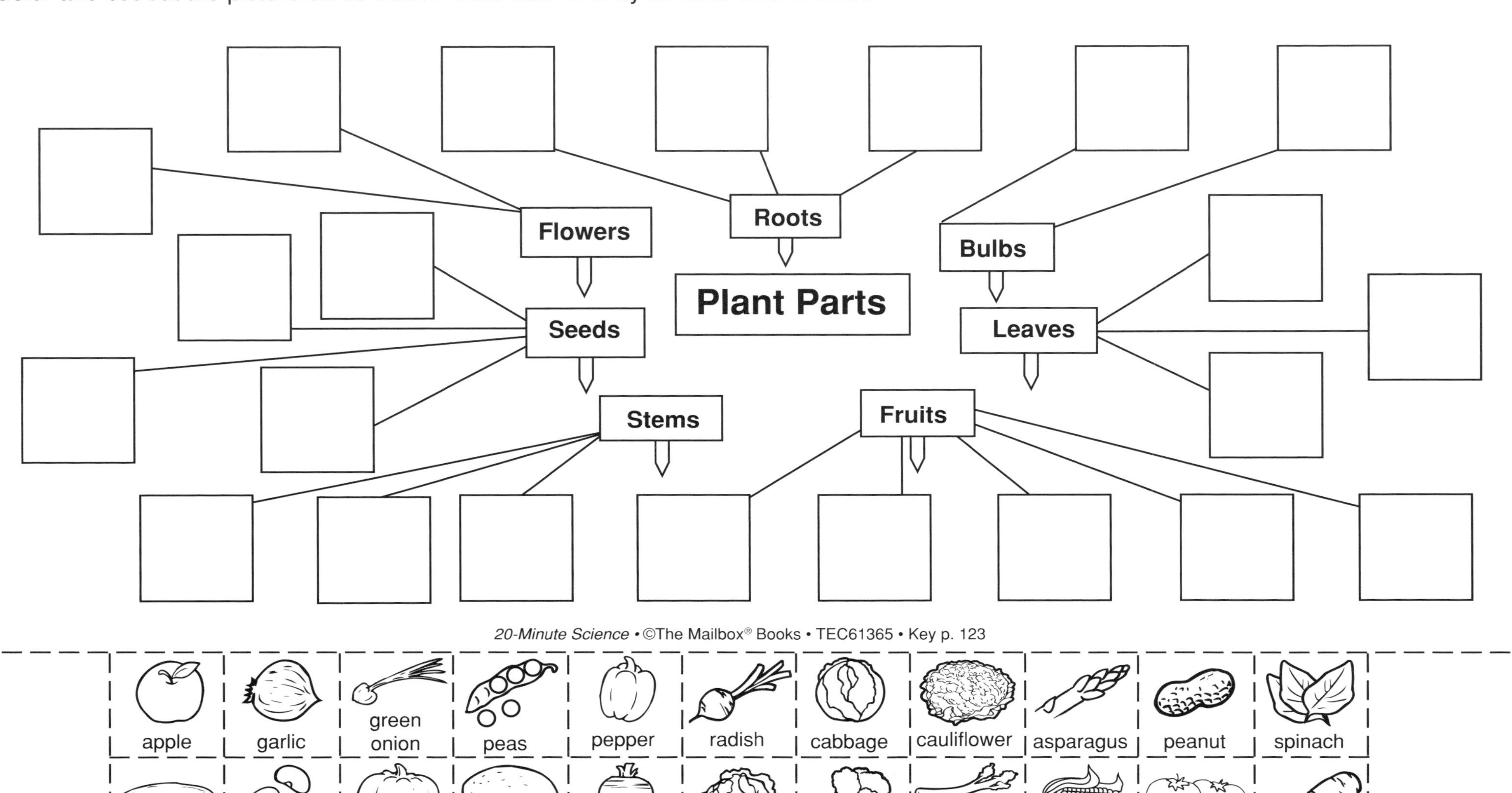

Note to the teacher: Use with “Edible Plant Parts” on page 21.

Dandelion Life Cycle Cards

Use with "Dandelion Life Cycle" on page 21.

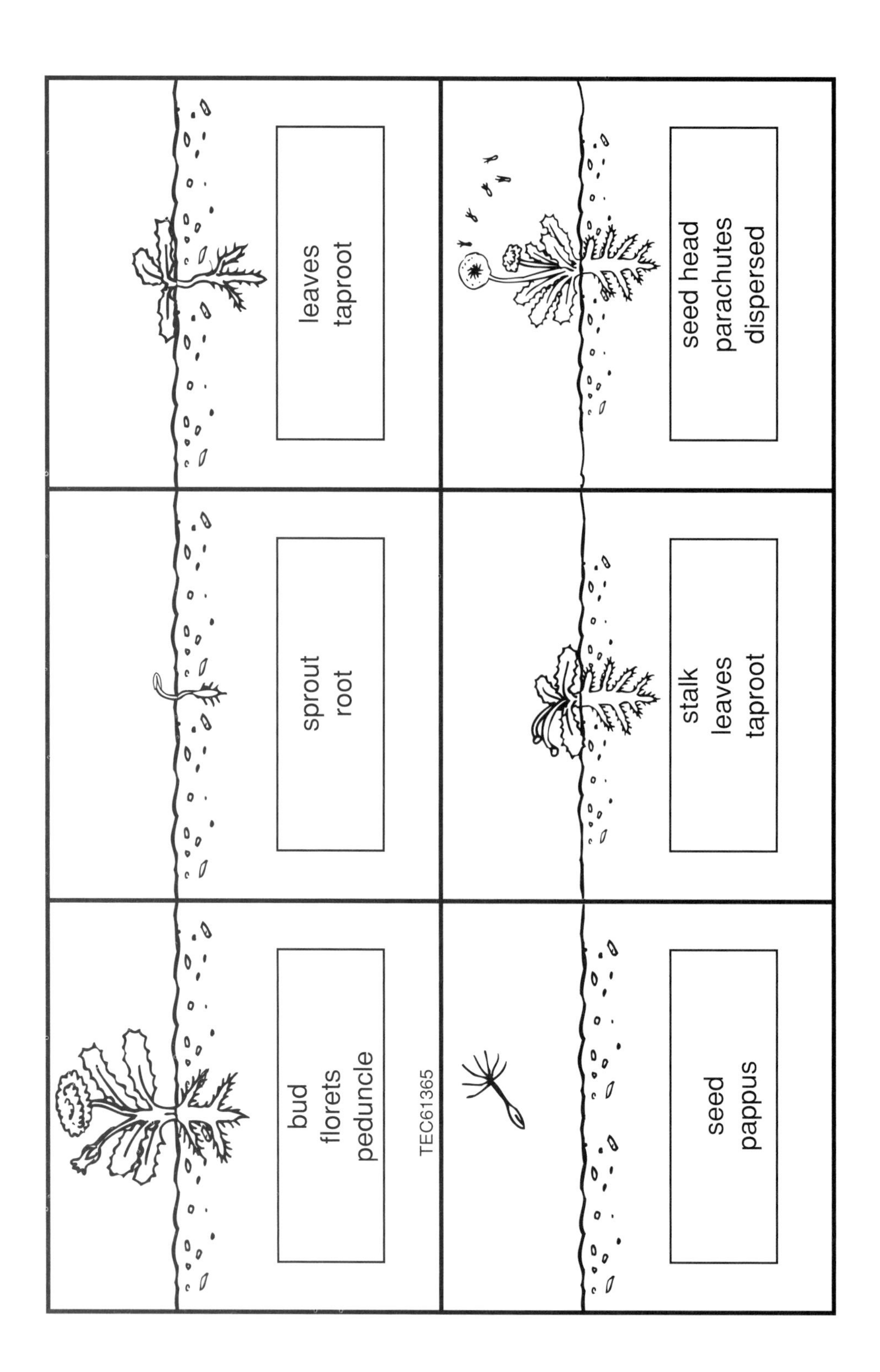

Name______________________________

Phenomenal Photosynthesis

Photosynthesis is the food-making process of green plants. To make their own food, plants need light, carbon dioxide, and water. A plant's leaves trap the energy from sunlight in a green substance called *chlorophyll* to create sugar (its food) and release oxygen (a waste product). Create a model of this food-making process of plants by following the directions below. Then answer the questions by writing a number in each blank.

Color Code
carbon (C) = orange
oxygen (O) = red
hydrogen (H) = blue

Directions:

1. Look at the four different shapes shown below: air bubbles, drops of water, a sugar cube, and clouds.
2. Inside each **bubble,** color each carbon atom and oxygen atom by the code. This creates six molecules of carbon dioxide.
3. Inside each **water drop,** color each hydrogen atom and oxygen atom by the code. This creates six water molecules.
4. On the **sugar cube,** color each carbon atom, oxygen atom, and hydrogen atom by the code. This creates one molecule of glucose.
5. Inside each **cloud shape,** color each oxygen atom by the code. This creates six molecules of oxygen.

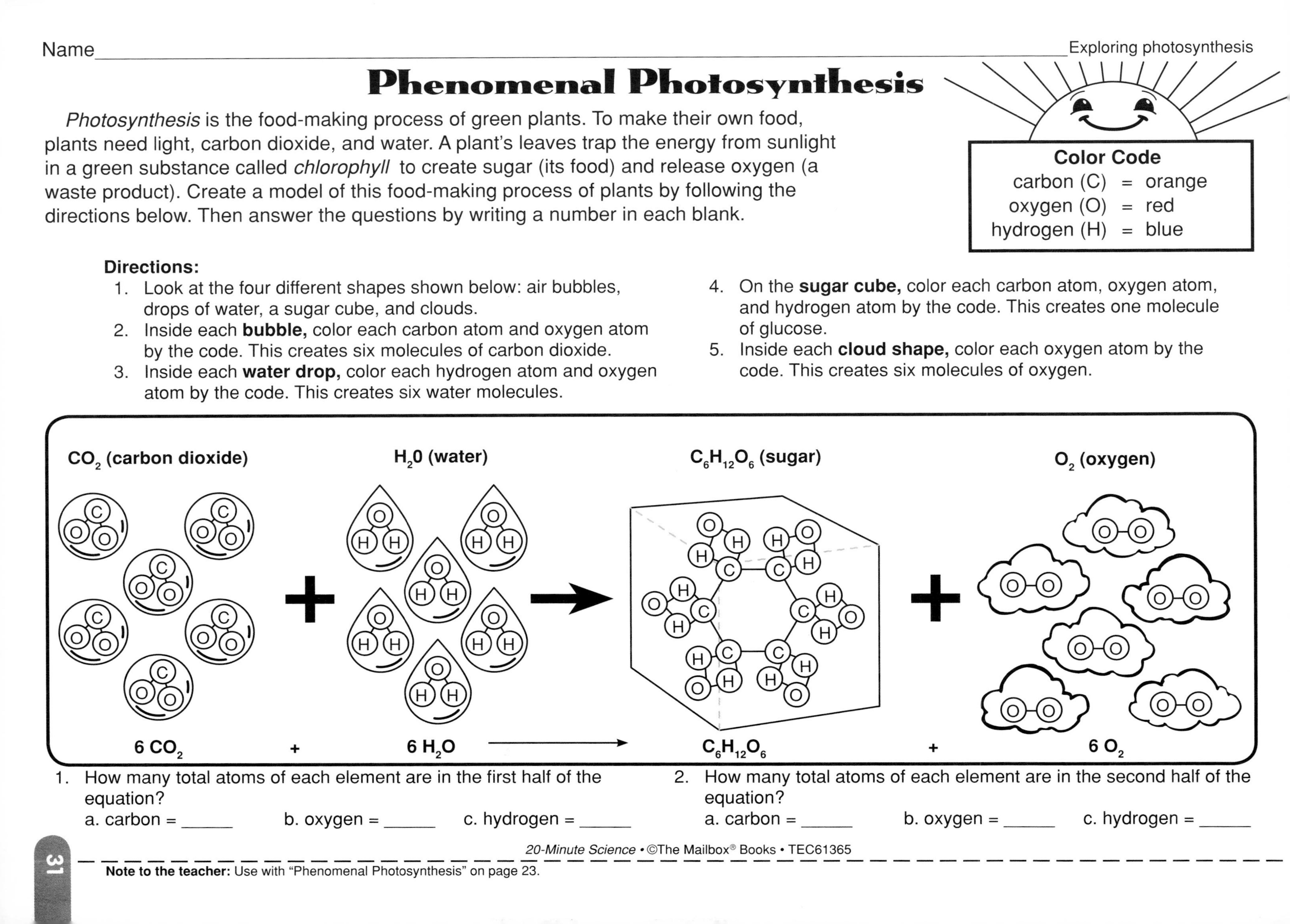

1. How many total atoms of each element are in the first half of the equation?
   a. carbon = _____ b. oxygen = _____ c. hydrogen = _____
2. How many total atoms of each element are in the second half of the equation?
   a. carbon = _____ b. oxygen = _____ c. hydrogen = _____

Note to the teacher: Use with "Phenomenal Photosynthesis" on page 23.

Name ______________________________ Identifying steps, completing analogies

Travelin' Along With Transpiration

Transpiration is the flow of water through a plant. But what path does the water take? Follow the directions in Part One and Part Two below to discover the water's path and learn more about transpiration and plants.

Part One: Complete each sentence below by writing a word from the word list in each blank. Also write the number of the matching sentence in each circle in the diagram. Use each word and number only once.

Word List

xylem vessels	stem
sun	absorbs
transpiration	cool
roots	stomata
evaporates	water

1. As the ________________ warms the water inside a plant's leaves, ________________ occurs. This warming makes much of the water change into water vapor that ____________.
2. The water vapor escapes into the atmosphere through the ______________________________ in the leaves.
3. The water vapor ________________ heat as it escapes, which helps the inside of the leaves ________________.
4. To replace the lost water, ________________ draw up more ________________.
5. It travels up the ________________ and along the veins of leaves through tiny tubes called ________________.

Part Two: Complete each analogy below by choosing the appropriate word(s) from the list provided. Use reference materials for help.

Word List

arteries and veins
perspiration
vitamins and minerals
pores
meat, cheese, and bread

1. Transpiration is to plants as ________________ is to humans.
2. Stomata are to plants as ________________ are to humans.
3. Water, carbon dioxide, and light are to photosynthesis as ______ ________________ are to a sandwich.
4. Xylem vessels are to plants as ________________ are to humans.
5. Nutrients are to plants as ________________ are to humans.

Bonus: On the back of this page, write an analogy of your own about plants.

Note to the teacher: Use with "Moving On Up!" on page 24.

Watch Out!

Did you know that there are about 700 kinds of poisonous plants in the United States and Canada? A poisonous plant is any plant that can injure a person or an animal. Many poisonous plants can be avoided because they look, taste, or smell disagreeable. *(Remember that even some familiar food plants can have poisonous parts!)*

Directions: Read each sentence below. Then use reference materials to help you fill in the spaces with the name of one of the poisonous plants shown. Finally, write the letter for each number in the secret code below.

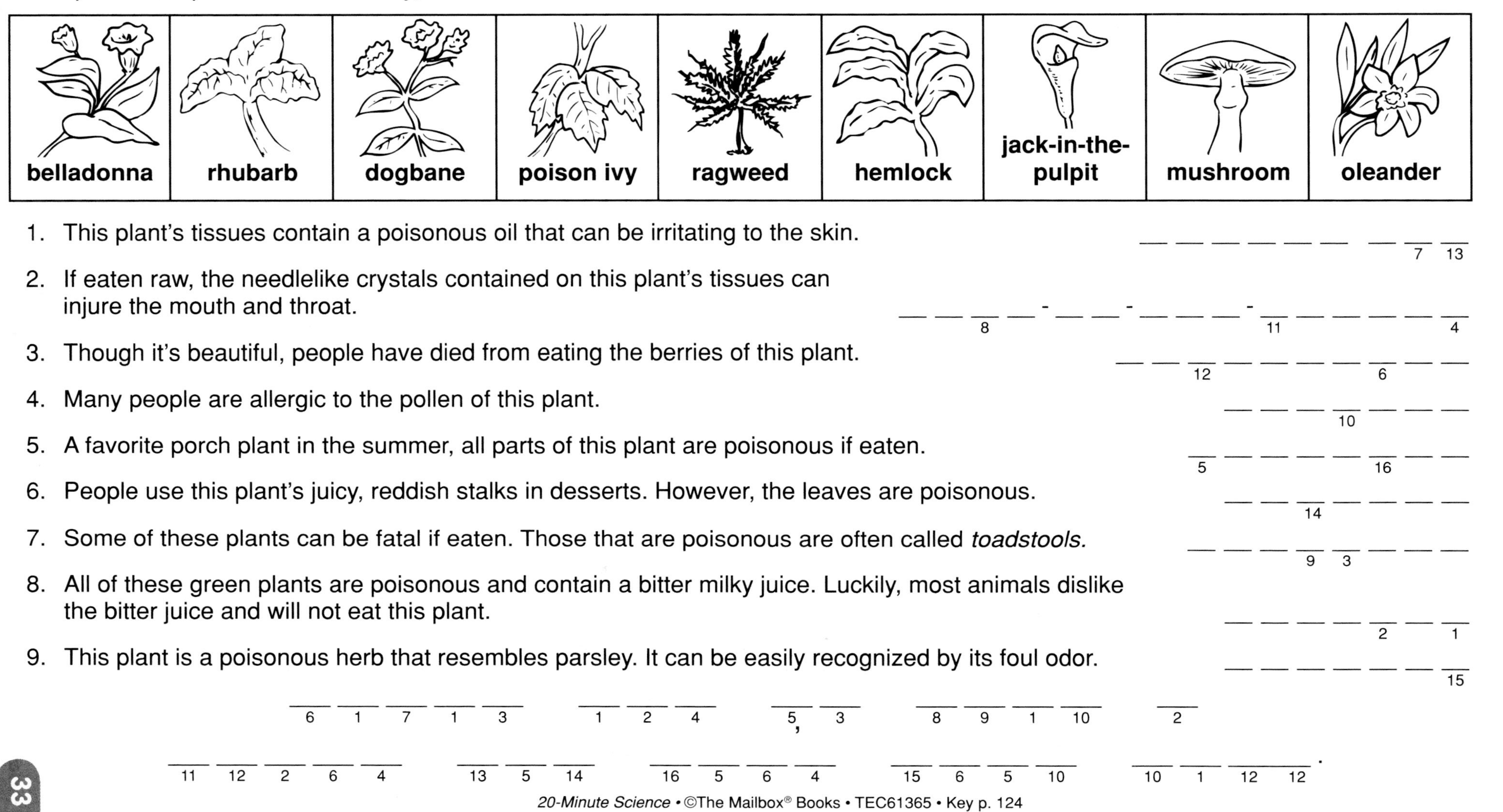

1. This plant's tissues contain a poisonous oil that can be irritating to the skin. ___ ___ ___ ___ ___ ___ ___ ___(7) ___(13)
2. If eaten raw, the needlelike crystals contained on this plant's tissues can injure the mouth and throat. ___ ___ ___(8) ___ - ___ ___ - ___ ___ ___ - ___(11) ___ ___ ___ ___ ___ ___(4)
3. Though it's beautiful, people have died from eating the berries of this plant. ___ ___ ___(12) ___ ___ ___ ___ ___(6) ___ ___
4. Many people are allergic to the pollen of this plant. ___ ___ ___ ___(10) ___ ___ ___
5. A favorite porch plant in the summer, all parts of this plant are poisonous if eaten. ___(5) ___ ___ ___ ___ ___(16) ___ ___
6. People use this plant's juicy, reddish stalks in desserts. However, the leaves are poisonous. ___ ___ ___(14) ___ ___ ___ ___
7. Some of these plants can be fatal if eaten. Those that are poisonous are often called *toadstools.* ___ ___ ___ ___(9) ___(3) ___ ___ ___
8. All of these green plants are poisonous and contain a bitter milky juice. Luckily, most animals dislike the bitter juice and will not eat this plant. ___ ___ ___ ___ ___(2) ___ ___(1)
9. This plant is a poisonous herb that resembles parsley. It can be easily recognized by its foul odor. ___ ___ ___ ___ ___ ___ ___(15)

6 1 7 1 3   1 2 4   5, 3   8 9 1 10   2

11 12 2 6 4   13 5 14   16 5 6 4   15 6 5 10   10 1 12 12.

Name ______________________________

Life Cycle Lingo

Read each definition. Write a letter from the grid that matches each ordered pair of symbols below the definition. Hint: To find each letter, read the grid across and then up.

1. the process of a new plant growing from part of an existing plant

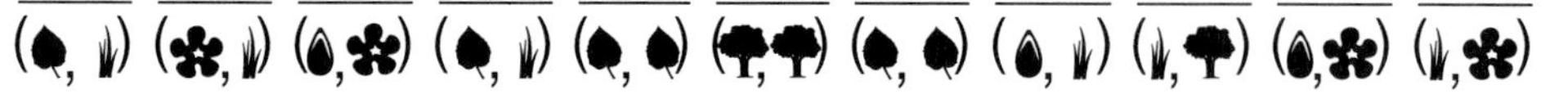

2. absorbed through a plant's leaves to make carbohydrates

3. develop in the flowering parts of plants and begin the next life cycle

4. used by plants to make protein

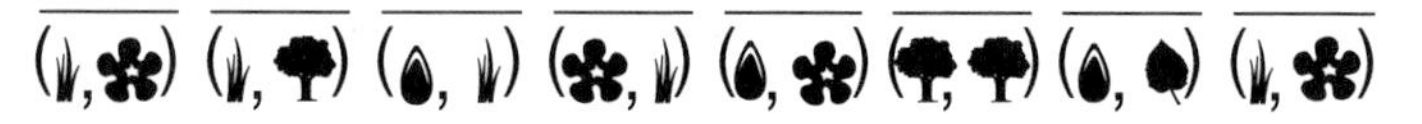

5. when a plant or plant community reaches full growth

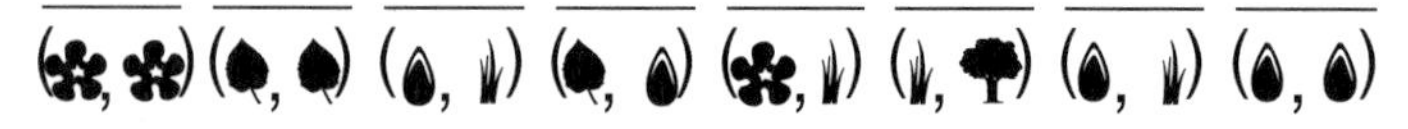

6. the sprouting or growth of a seed

7. brought through the soil by water to help plants grow

8. the natural replacing of one plant community for another

9. the spreading or scattering of seeds

10. to rot or decay

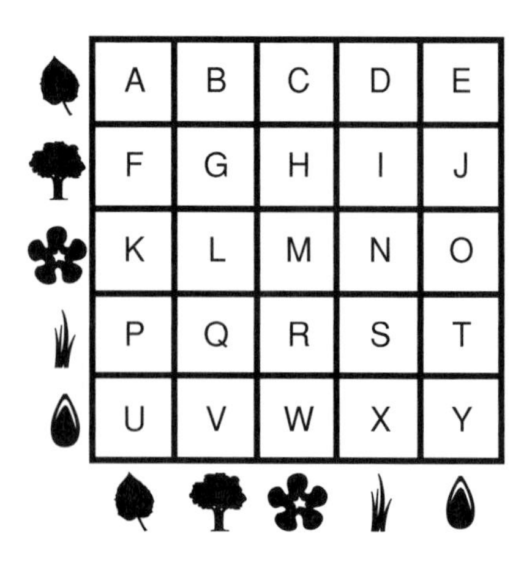

A	B	C	D	E
F	G	H	I	J
K	L	M	N	O
P	Q	R	S	T
U	V	W	X	Y

Respiration and Photosynthesis

Carefully read the paragraphs below to learn how dependent people and plants are on one another for oxygen and carbon dioxide. Then, in the outline on the left, explain why the exchange of these gases is important to people. In the outline on the right, explain why the exchange of these gases is important to plants. Use your own words to write your explanations.

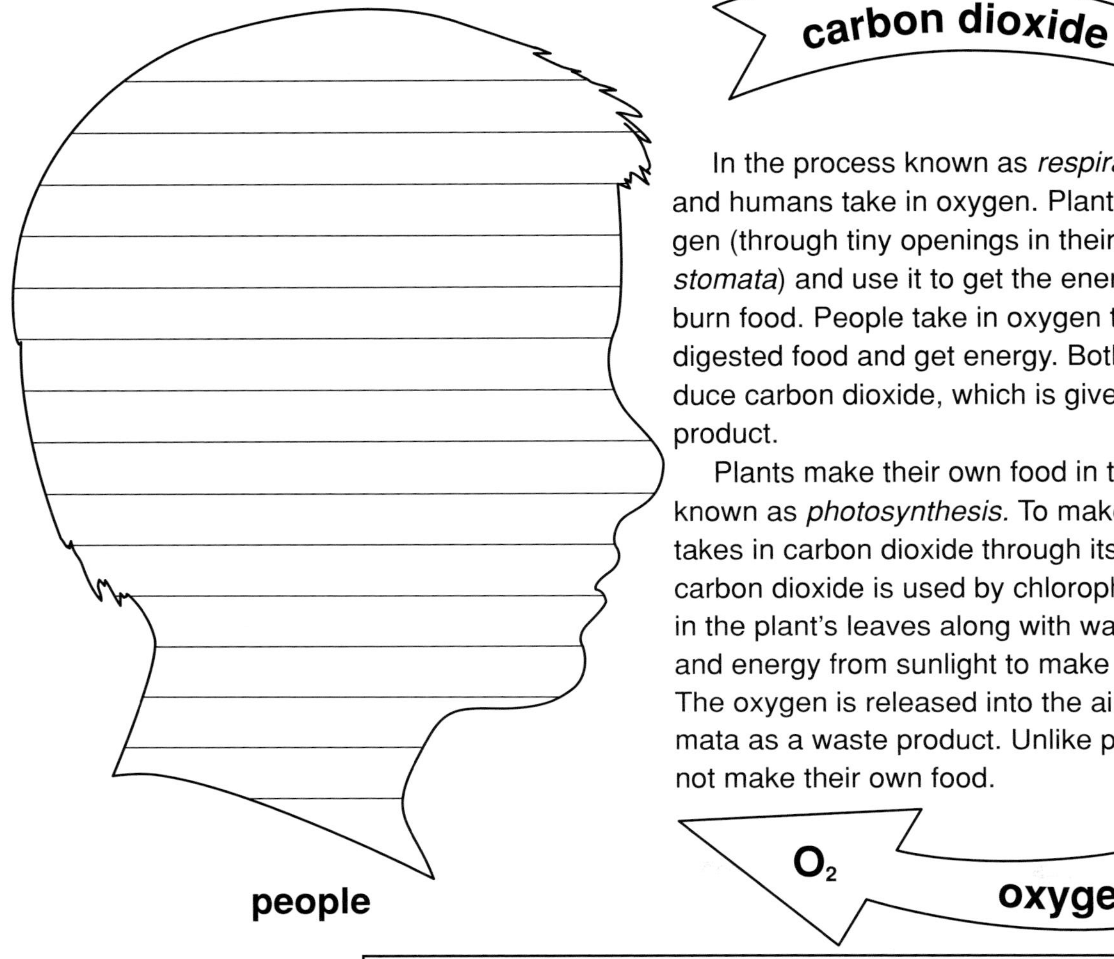

In the process known as *respiration,* both plants and humans take in oxygen. Plants take in oxygen (through tiny openings in their leaves called *stomata*) and use it to get the energy they need to burn food. People take in oxygen to help them burn digested food and get energy. Both processes produce carbon dioxide, which is given off as a waste product.

Plants make their own food in the process known as *photosynthesis.* To make food, a plant takes in carbon dioxide through its stomata. The carbon dioxide is used by chlorophyll (green matter) in the plant's leaves along with water from its roots and energy from sunlight to make food and oxygen. The oxygen is released into the air through the stomata as a waste product. Unlike plants, people cannot make their own food.

Bonus: Yeast and bacteria use anaerobic respiration. Use a dictionary to learn the meaning of *anaerobic*. Then write a sentence on the back of this page explaining what *anaerobic respiration* is.

Name ______________________________ Identifying plants' needs

Oops! What's Missing?

The scientists at Essential Plant Needs, Inc., know that land plants need light, water, soil, and air to be healthy. Help these scientists keep up the good work by identifying which important element (light, water, soil, or air) is missing for the plant in each situation below. *Hint: Some may be missing more than one element.*

Bonus: Illustrate one of the cases above on the back of this paper.

Animals

Variety of Vertebrates

Materials for each group:
5" x 25" strip of colorful paper
access to reference materials

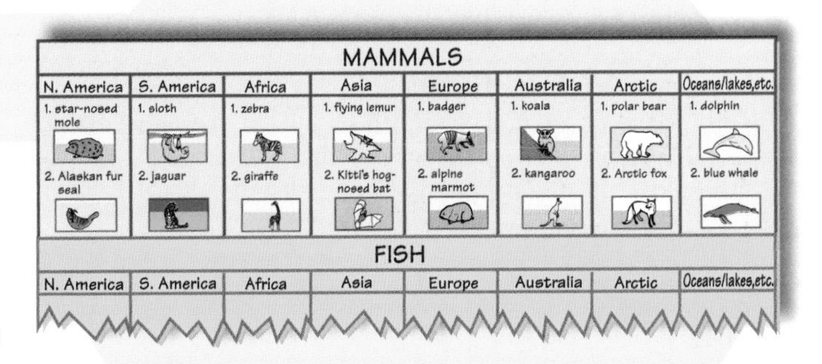

Share with students characteristics of the following five classes of vertebrates: mammals, birds, amphibians, reptiles, and fish. Explain that vertebrates can be found in many parts of the world. Then divide students into five groups and assign each group a different vertebrate class. Set aside two 20-minute sessions for research.

Research: Instruct each group to make a chart similar to the one shown. Then challenge each group to find two vertebrates in its assigned class for each category on the chart. Have them list each vertebrate in the appropriate column and add an illustration.

Follow-up: Invite each group to share what it learned about its assigned vertebrate class. Then display each group's chart, one above the other, to make a large reference chart.

Did You Know?
There are about 40,000 species of vertebrates.

Collect a Class

Students will learn some of the animals in each of the five most common vertebrate classes when they play this fun game, similar to the traditional game of Go Fish. Before playing, divide students into groups of three to five players.

Materials for each group:
copy of page 42
access to reference materials

Steps for students:
1. Fill in the answer box by using reference materials to list each animal under its correct class.
2. Cut apart and shuffle the animal cards. Deal five cards to each player. Stack the remaining cards facedown in the playing area.
3. Take turns asking the student on your left for one of the following: a reptile, a fish, an amphibian, a mammal, or a bird card. If that player doesn't have a card from that particular class, draw a card from the center pile.
4. Play continues until one player has all five cards from one class. (Use the answer box to check the cards.)

Incredible Invertebrates

Have students make these review cards to help them characterize the eight major groups of invertebrates. On the board, list each phylum shown. Share the description of each phylum and its examples with students. Then lead students in the steps below to create a set of invertebrate review cards.

Phylum	Description	Examples
Porifera	pore-bearing animals	sponges
Coelenterates	animals with special stinging organs	jellyfish, coral sea anemone
Platyhelminthes	worms with flattened bodies	planarian, fluke, tapeworm
Nematoda	worms with rounded bodies	pinworm, trichina hookworm
Annelida	worms with segmented bodies	leech, earthworm, sandworm
Echinodermata	animals with external spines	starfish, sea urchin, sand dollar
Mollusca	soft-bodied animals, usually with hard outer shells	clam, squid, octopus
Arthropoda	animals with jointed legs and exoskeletons, often like armor	lobster, insect, spider

Annelida

6

worms with segmented bodies

earthworm

Materials for each student:

copy of the invertebrate cards on the top half of page 43
copy of the skill sheet on page 44
access to reference materials
8 index cards
access to a hole puncher
metal loose-leaf ring

Day 1:
Steps for students:

1. Copy a different phylum from the list on the front of each card.
2. Cut out the cards from page 43. Glue each card on the top half of the back of the corresponding phylum card.

Day 2:
Steps for students:

1. Use the reference materials to draw an animal from each phylum on the back of the appropriate card.
2. Hole-punch the upper left corner of each card and secure the cards on the metal ring.

Day 3: Have students use their cards to help them complete a copy of page 44.

Adaptations

teeth—incisors, molars, canines
mouthparts—tube, beak
movement—wings, powerful arms and legs
coverings—moist skin, hair, armor
coloration—color is similar to surroundings
resemblance—looks similar to something in surroundings
mimicry—looks like a dangerous or poisonous animal
defenses—escaping, fighting
nest building—building different types of nests
hibernation—deep sleep
migration—moving back and forth from region to region
practice—repeating the process
reinforcement—punishment or reward given
imprinting—learning caused by one experience that establishes a pattern

Need to Adapt?

Materials:

chart like the one shown
class supply of 9" x 12" white construction paper
access to reference materials

Day 1: This activity helps students understand the many types of animal adaptations. To begin, explain to students that animals adapt to changes in their environment in order to survive. As an example, tell students that a snowshoe hare changes color with the seasons so that it can blend in with its environment. Share with students the chart of animal adaptations shown. Then assign each child an adaptation to research.

Day 2: Have each child research his assigned adaptation to find several animals that have that particular adaptation. Then have each child make a mini poster to highlight the adaptation.

Building a Biome

To begin, explain to students that a biome is a plant and animal community that covers a large geographic area. Then discuss with students the different examples of land and aquatic biomes shown in the box.

Materials for each pair of students:
shoebox
access to reference materials
access to magazines
access to materials such as plastic figurines, grass, twigs, and cotton balls

Day 1: Pair students. Ask each pair to select a biome and investigate the climate, geographic features, plants, and animals of its selected biome, plus one or more places where that biome can be found.

Days 2 and 3: Have each twosome use a variety of materials to create a diorama showing the specific features of its chosen biome.

Day 4: Invite each pair, in turn, to share with the class its diorama and accompanying research.

Land Biomes	Aquatic Biomes
Tundra	Ponds and Lakes
Taiga	Streams and Rivers
Temperate Coniferous Forest	Wetlands
Temperate Deciduous Forest	Oceans
Chaparral	Coral Reefs
Desert	Estuaries
Grassland	
Savanna	
Tropical Rain Forest	
Tropical Seasonal Forest	

Name Becca

Exploring habitats

A Butterfly 's Habitat

Description: The butterfly was on a flower until a bird chased it. Then the butterfly flew to a dead log where it rested for a while. Its colors helped it blend with the surroundings.

A "Hand-y" Habitat

Help students get better acquainted with plant and animal habitats.

Materials for each student:
copy of page 45
access to reference materials

Day 1: Review with students that a habitat is a place that provides any plant or animal with food, water, shelter, and space to live. Have each child choose a different plant or animal to research and begin to investigate her plant or animal's habitat.

Day 2: Have each child continue to investigate her plant or animal's habitat. Direct each student to illustrate five important facts about her topic's habitat, each on a different finger and on the thumb, on her copy of page 45. Then have her write a description of her topic's habitat on the palm.

Adaptation Lotto

Materials for each student:
copy of the lotto gameboard on page 43
nine game markers or counters

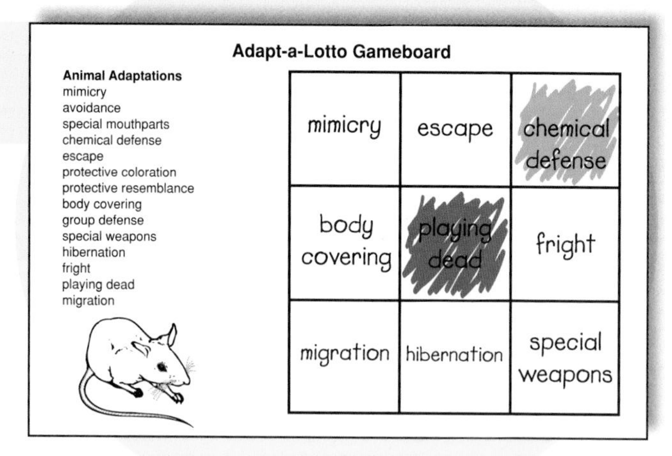

Add adaptations to a traditional game of lotto for a quick review. To prepare, have each child write nine adaptations from the list on her gameboard, each in a separate box. To play, read aloud one of the adaptation clues shown. Direct any student whose gameboard has the adaptation being described to cover it with a marker. Continue playing in this manner until one student has covered a vertical, horizontal, or diagonal row and calls out "Adapt-a-Lotto!" If desired, play again, directing students to fill two horizontal or vertical rows, the four corners, or the entire gameboard to win.

Clues

- Salamanders stay underground during cold weather. *(hibernation)*
- Gray whales spend the summer in the Bering Sea and the fall in Baja California. *(migration)*
- A frilled lizard raises its special folds of skin, making an attacker think it is larger. *(fright)*
- The viceroy butterfly is often mistaken for the poisonous monarch butterfly. *(mimicry)*
- A walking stick looks a lot like the branches on which it lives. *(protective resemblance)*
- The chameleon turns green when it is on green leaves. *(protective coloration)*
- The Goliath beetle can't run fast, but it has a thick armor. *(body covering)*
- The butterfly has a proboscis, shaped like a drinking straw, which helps it get nectar from flowers. *(special mouthparts)*
- A skunk squirts a horrible-smelling liquid on its attacker. *(chemical defense)*
- Caribou have antlers that are large enough to fend off grizzly bears. *(special weapons)*
- A gibbon stays off the ground. It spends its entire life in the trees. *(avoidance)*
- An impala can run at a speed of 50 miles per hour. *(escape)*
- Musk oxen defend their calves by standing in a circle with their horns facing outward. *(group defense)*
- A North American opossum closes its eyes and becomes totally limp in the presence of an enemy. *(playing dead)*

Life Science

Creature Features

Materials:
chart paper labeled with the list of animals shown
blank card for each student
access to reference materials

Challenge students to identify animals based on their adaptations with a fun game. First, assign each child a different animal from the list provided. Then direct each student to write three to five sentences about his animal's adaptations (without using his animal's name) on a blank card, using reference materials as needed.

Follow-up: Direct several students (one at a time) to read their cards aloud. Then invite the group to guess each featured animal.

1. It can stay underwater for three or four minutes.
2. This animal is an expert swimmer.
3. It has a layer of fat that insulates it from the cold.
4. It can use its paws to handle objects, such as stones and small shellfish.
5. It has elastic webbing between its toes.

Animals

bighorn sheep	otter	roadrunner	killer whale
golden eagle	spotted owl	scorpion	manatee
chinchilla	ocelot	Arctic hare	octopus
giraffe	spider monkey	polar bear	purple sea urchin
ostrich	two-toed sloth	musk ox	jellyfish
eland	tarantula	snowy owl	
aardvark	driver ant	emperor penguin	
moose	dingo		

Decomposers at Work

Materials:
copy of page 46 for each student
clear quart-size jar with screw-on lid
a few small food scraps (bread, fresh fruit or vegetables, or cheese—no meat or fish)
masking tape
small hand towel

Day 1: Tell students that decomposers (organisms such as bacteria and fungi) feed on dead plants and animals. Explain that you are going to construct a habitat for decomposers. Then put the food scraps in the jar and tighten the lid. Wrap masking tape around the area between the jar and the lid. Fold the towel and place the jar on its side atop the folded towel (to keep the jar from rolling). Have each child list on her copy of page 46 the foods you placed in the jar.

Observation: Every other day (Monday, Wednesday, Friday) for two weeks, have each child record on her copy of page 46 the date and a description of the jar's contents.

Follow-up: After the final observation period, direct students to answer the questions at the bottom of page 46. Then lead students in a discussion of the results.

See the animal skill sheets on pages 47–49.

Name ______________________________

Collect a Class

Follow your teacher's directions.

puffin	Komodo dragon	bush baby	salamander	stingray
tadpole	hedgehog	peacock	shark	gecko
gharial	sea horse	arrow-poison frog	platypus	owl
flamingo	spiny anteater	python	Gila monster	mudskipper
eel	giant tortoise	caecilian	dolphin	emu

Answer Box

mammals	fish	reptiles	amphibians	birds

Note to the teacher: Use with "Collect a Class" on page 37.

Invertebrate Cards

Use with "Incredible Invertebrates" on page 38.

pore-bearing animals	animals with special stinging organs
worms with flattened bodies	worms with rounded bodies
worms with segmented bodies	animals with external spines
soft-bodied animals, usually with hard outer shells	animals with jointed legs and exoskeletons, often like armor

Adapt-a-Lotto Gameboard

Animal Adaptations
mimicry
avoidance
special mouthparts
chemical defense
escape
protective coloration
protective resemblance
body covering
group defense
special weapons
hibernation
fright
playing dead
migration

Note to the teacher: Use with "Adaptation Lotto" on page 40.

Name ____________________

Invertebrates in Action

Stephen Spineberg, a famous film director, is having auditions for parts in *The Empire Strikes Backless,* the upcoming sequel to his successful movie *Spine Wars.* Stephen wants to fill each role with the actor whose attributes best match those of the character being cast.

Read the descriptions below. Decide which actor is best suited for each role. Write his or her name on the line provided.

Actors

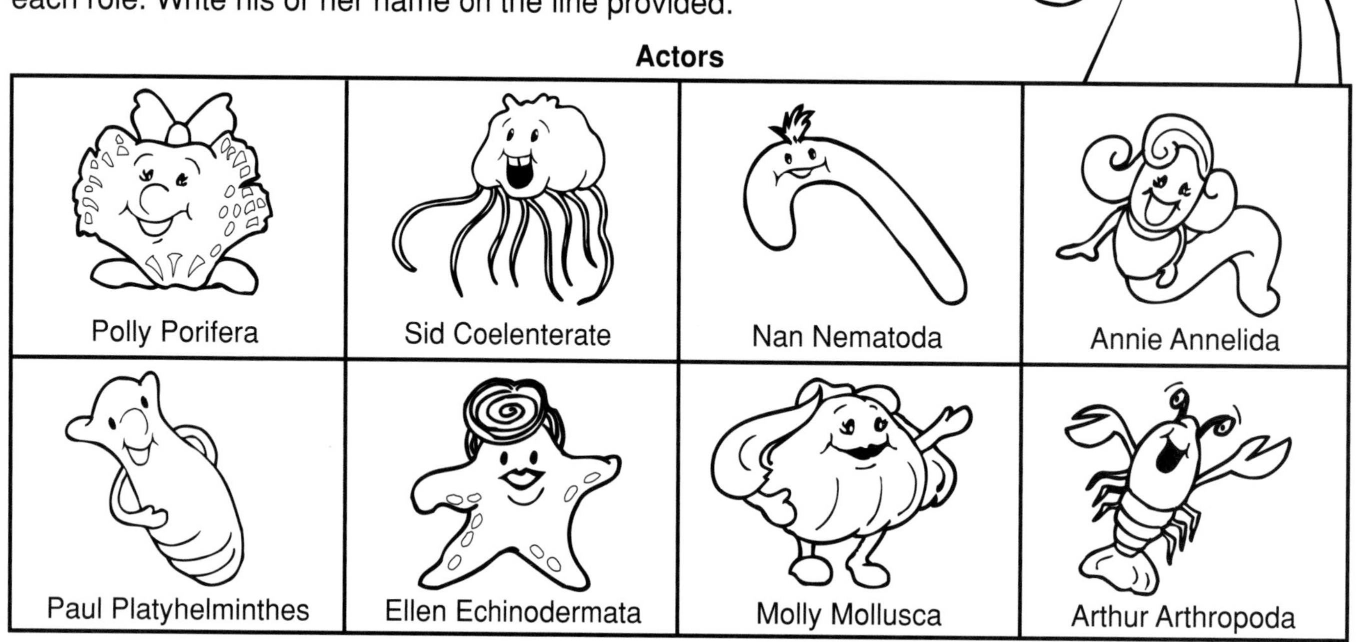

Role 1: ____________________
This role must be played by an animal that has an external spine and radial symmetry.

Role 2: ____________________
This role must be played by a soft-bodied animal that has a hard outer shell.

Role 3: ____________________
This role must be played by an animal that has jointed legs and an exoskeleton of chitin.

Role 4: ____________________
This role must be played by a worm that has a segmented body.

Role 5: ____________________
This role must be played by a worm that has a flattened body.

Role 6: ____________________
This role must be played by a sponge, or pore-bearing animal.

Role 7: ____________________
This role must be played by an animal that lives in the ocean and has a special stinging organ called a nematocyst.

Role 8: ____________________
This role must be played by a roundworm, preferably a pinworm or a hookworm.

Note to the teacher: Use with "Incredible Invertebrates" on page 38.

Name ____________________

A ____________________'s

Habitat

Description: ____________________

____

____

____

____

____

____

____

____

Note to the teacher: Use with "A 'Hand-y' Habitat" on page 39.

Name ______________________

Decomposers at Work

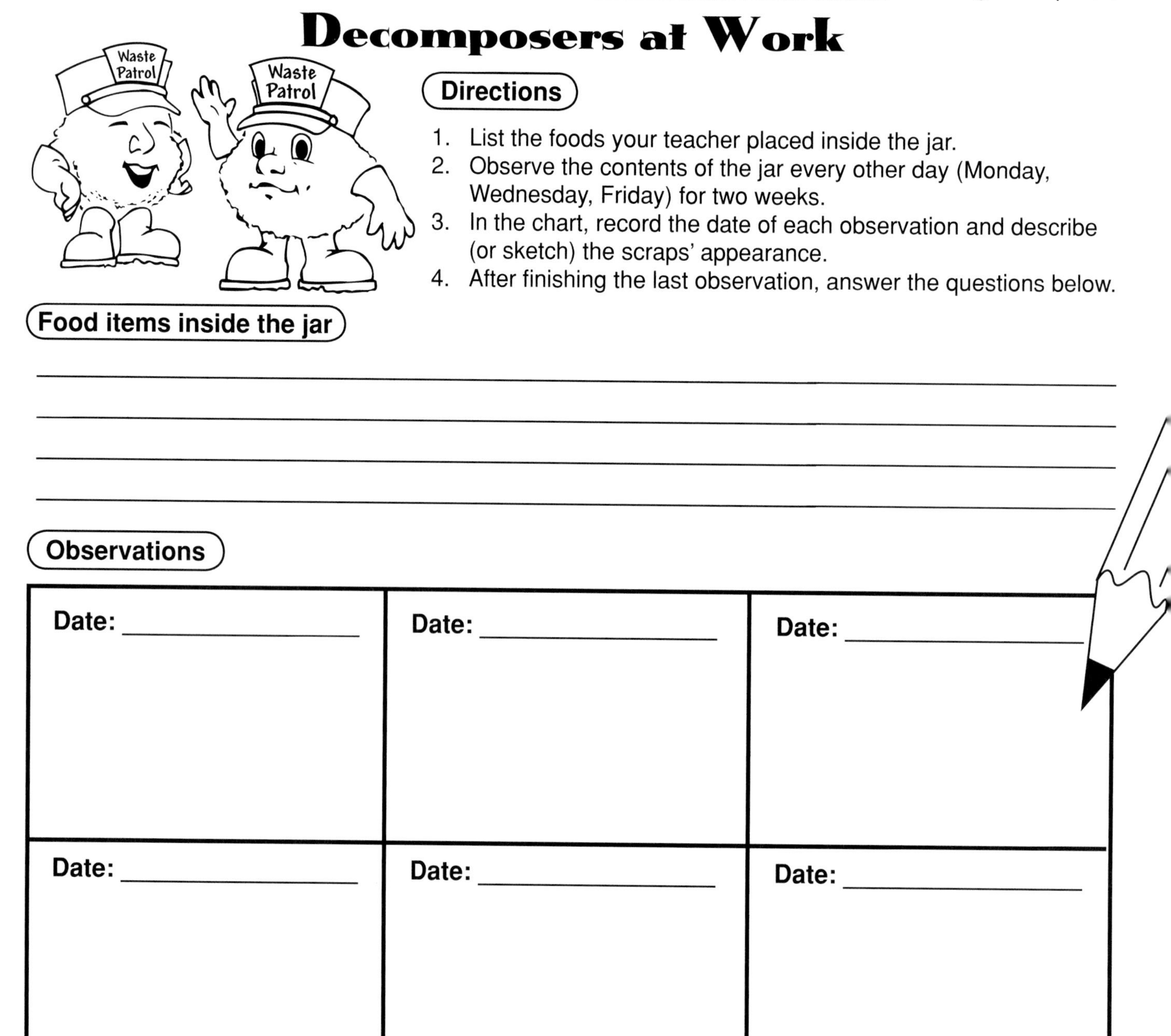

Directions

1. List the foods your teacher placed inside the jar.
2. Observe the contents of the jar every other day (Monday, Wednesday, Friday) for two weeks.
3. In the chart, record the date of each observation and describe (or sketch) the scraps' appearance.
4. After finishing the last observation, answer the questions below.

Food items inside the jar

____

____

____

____

Observations

Date: ________	Date: ________	Date: ________
Date: ________	Date: ________	Date: ________

Questions

1. How could you tell that decomposition was taking place? ______________________

____

2. How do you think the decomposers got into the jar? ______________________

____

____

3. If the scraps were to remain in the jar for two more weeks, how do you think they would look? ______________________

____

Note to the teacher: Use with "Decomposers at Work" on page 41.

Name ______________________

Habitat Hunt

Agent Aardvark is on a hunt to determine how different animals protect themselves in their habitats. Help the agent assemble his data by following the directions below.

Directions: Use a ruler to draw a line to connect each animal to its habitat. On the lines below, write a phrase explaining the adaptations the animal has in order to live in its habitat. Then answer the riddle at the bottom of the page by writing the letters that the lines cross in the correct spaces.

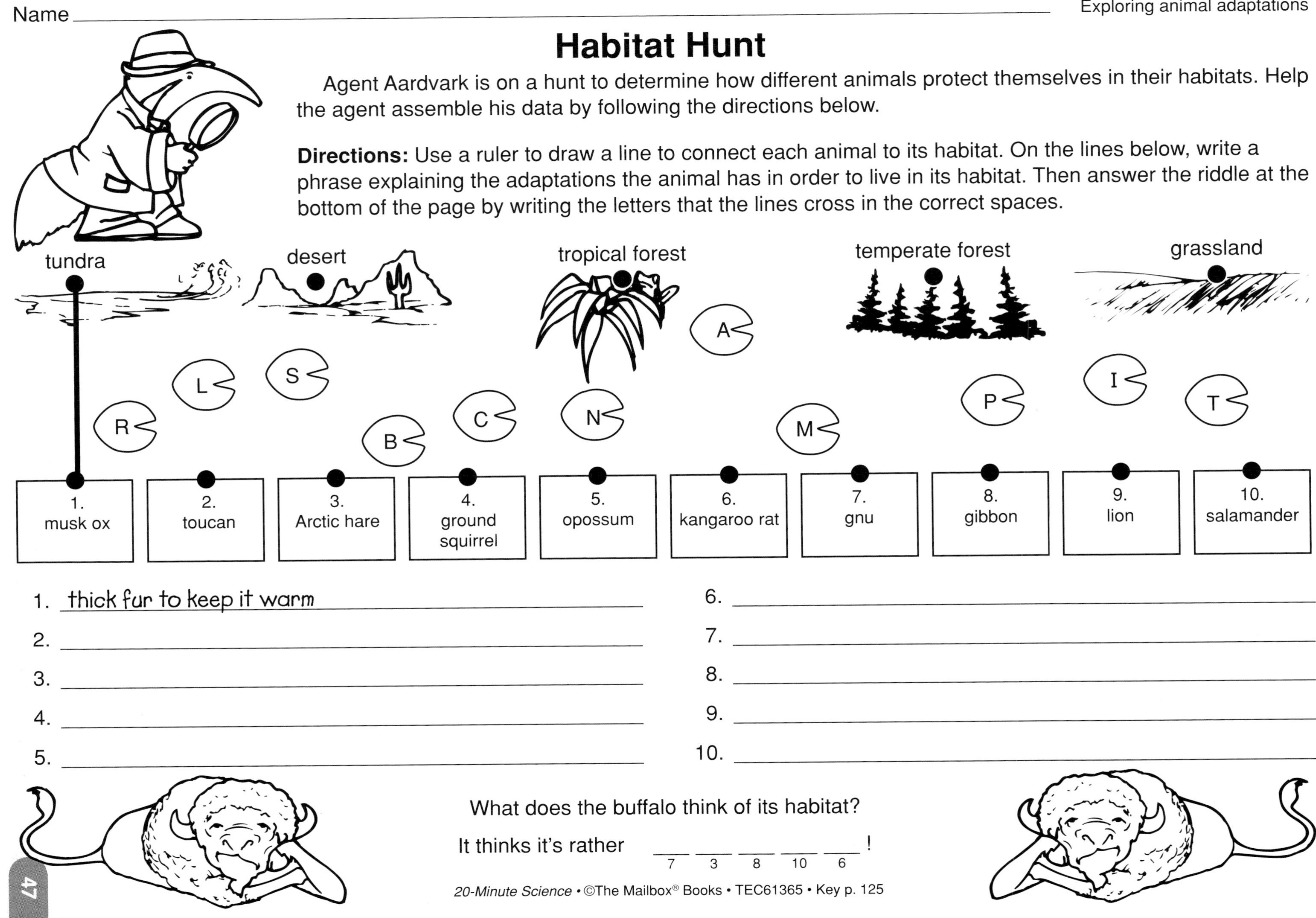

1. thick fur to keep it warm
2. ______________________
3. ______________________
4. ______________________
5. ______________________
6. ______________________
7. ______________________
8. ______________________
9. ______________________
10. ______________________

What does the buffalo think of its habitat?

It thinks it's rather ___ ___ ___ ___ ___!
(7 3 8 10 6)

Name ____

Surfing the Wildlife Web!

Professor Wyle D. Animal is surfing the Web for sites with never-before-seen wildlife. Follow the directions below to design a Web page that will really get the professor's attention.

Directions:

1. Choose a habitat from the list below.
2. Using the questions as a guide, create a new animal that could survive in the habitat you've chosen. Then draw the new animal on the computer screen at the bottom of the page.
3. Enhance your Web page by adding words and illustrations that tell about your animal and its habitat.

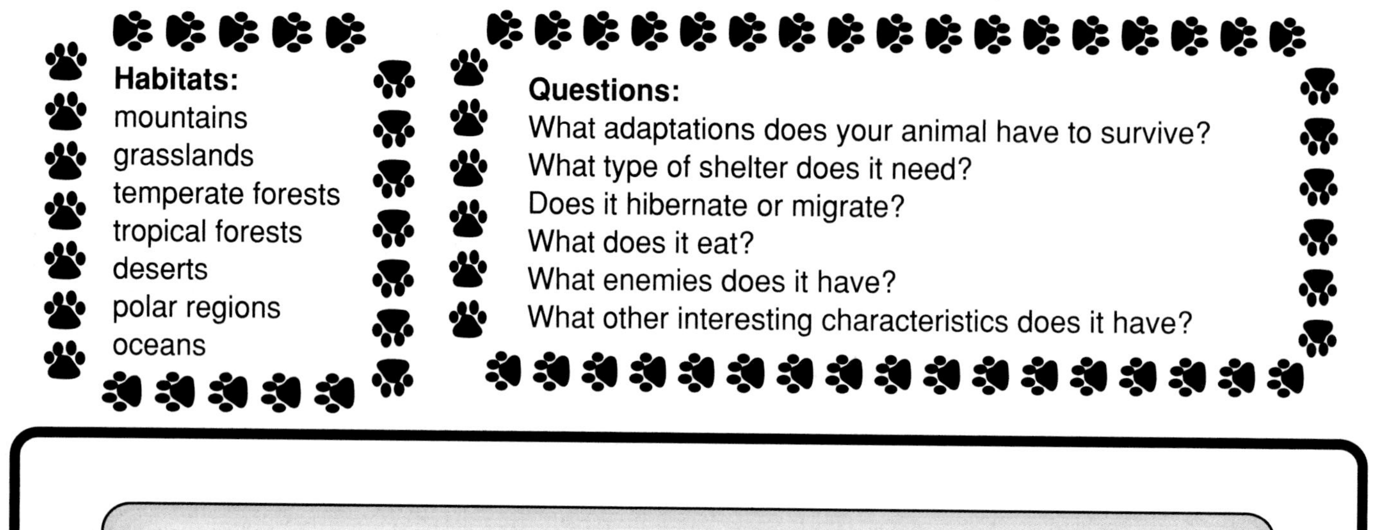

Habitats:
mountains
grasslands
temperate forests
tropical forests
deserts
polar regions
oceans

Questions:
What adaptations does your animal have to survive?
What type of shelter does it need?
Does it hibernate or migrate?
What does it eat?
What enemies does it have?
What other interesting characteristics does it have?

Address: www. ____________________________ .com

____
title

Name ______________________________

All About the Food Chain

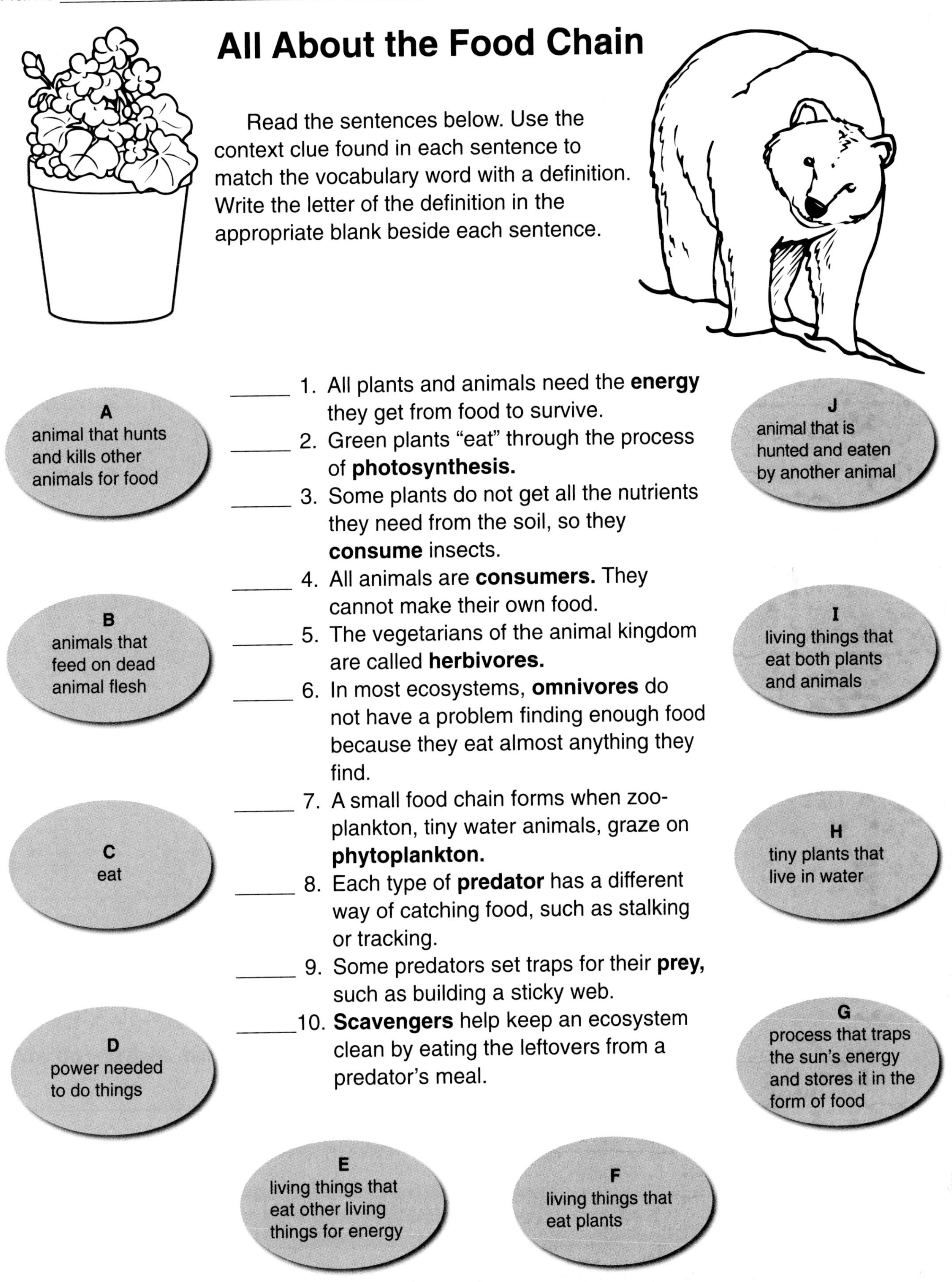

Read the sentences below. Use the context clue found in each sentence to match the vocabulary word with a definition. Write the letter of the definition in the appropriate blank beside each sentence.

_____ 1. All plants and animals need the **energy** they get from food to survive.

_____ 2. Green plants "eat" through the process of **photosynthesis.**

_____ 3. Some plants do not get all the nutrients they need from the soil, so they **consume** insects.

_____ 4. All animals are **consumers.** They cannot make their own food.

_____ 5. The vegetarians of the animal kingdom are called **herbivores.**

_____ 6. In most ecosystems, **omnivores** do not have a problem finding enough food because they eat almost anything they find.

_____ 7. A small food chain forms when zoo-plankton, tiny water animals, graze on **phytoplankton.**

_____ 8. Each type of **predator** has a different way of catching food, such as stalking or tracking.

_____ 9. Some predators set traps for their **prey,** such as building a sticky web.

_____ 10. **Scavengers** help keep an ecosystem clean by eating the leftovers from a predator's meal.

A animal that hunts and kills other animals for food

B animals that feed on dead animal flesh

C eat

D power needed to do things

E living things that eat other living things for energy

F living things that eat plants

G process that traps the sun's energy and stores it in the form of food

H tiny plants that live in water

I living things that eat both plants and animals

J animal that is hunted and eaten by another animal

Bouncing Light

Materials:
rubber ball

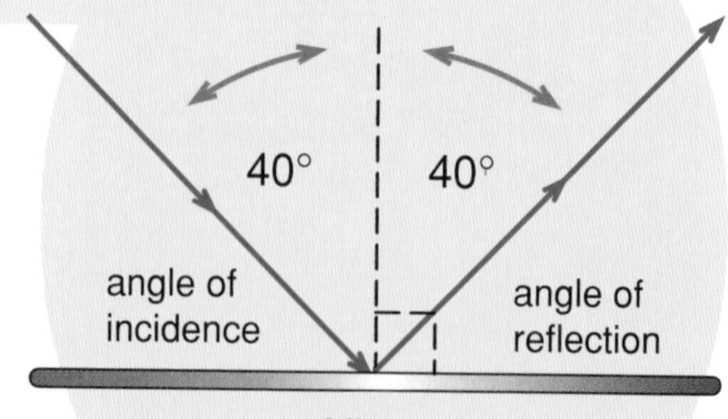

Help students gain an understanding of reflection with this simple demonstration. Invite two volunteers to stand in front of the class. Direct Student 1 to bounce the ball at an angle to Student 2. Ask the class to describe the ball's bounce *(bounces at an angle from Student A to Student B).* Instruct Students A and B to bounce the ball between them at different angles. Tell the class that the way the ball bounces (in a straight line) is how light bounces, or *reflects*, from mirrors and other flat, shiny surfaces.

Next, copy the drawing shown onto the board. Explain that the angle of incidence (the angle at which light from an object hits the mirror) equals the angle of reflection (the angle at which light rays from the object are reflected). Point out that this is how people are able to see images in a mirror even when they cannot see themselves in it—because they are standing in the angle of reflection.

Did You Know?
Mirrors reflect (bounce back) most of the light that strikes them.

Reflecting Light

Materials:
flashlight
4 small mirrors

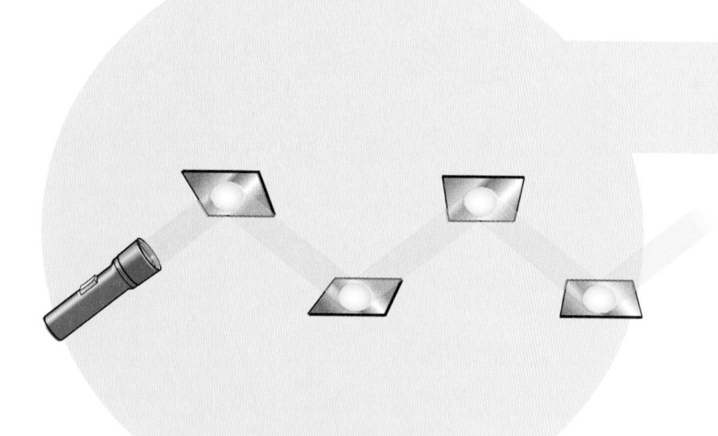

Invite students to observe the characteristics of reflected light by having them make light beams zigzag through the classroom. Give each of four children a mirror. Ask these students to hold their mirrors waist high and stand facing each other in two parallel lines about three feet apart. Darken the room and have a fifth child shine the flashlight at the mirror of the first child. Direct that child to use his mirror to reflect the light to the mirror—not the eyes—of the child across from him. Have that child reflect the light to the child across from him. Continue until the light reaches the last child. Repeat the demonstration with different students until each child has had a chance to participate.

Off the Wall

In advance, prepare paper targets by numbering five sheets of dark-colored construction paper; then mount them on a wall in five separate locations, each level with a desktop. Direct groups of students to follow the steps below.

Materials for each group of students:
small mirror
flashlight
plain sheet of paper
tape
ruler
protractor

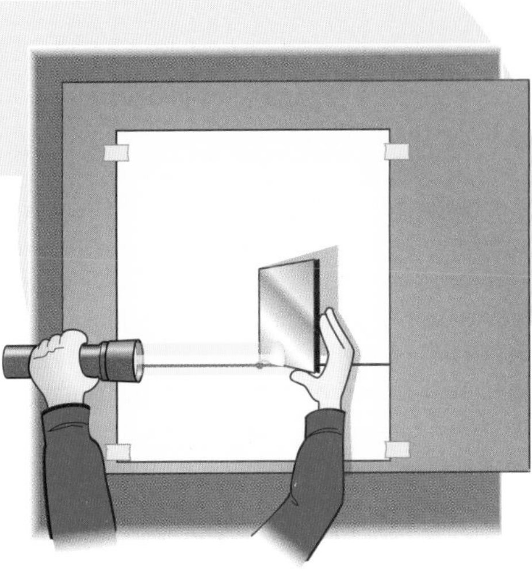

Day 1:
Steps for students:

1. Tape the paper to a desktop near a paper target. Use the ruler to draw a horizontal line across the paper.
2. Make a dot (the vertex of the angles to be drawn) anywhere along the line.
3. Place the flashlight on the desktop, aligning it with the line on the left side of the paper.
4. Place the bottom left corner of the mirror on the dot and turn the flashlight on. Darken the room, if necessary. Pivot the mirror (not the flashlight or paper) on the dot until the reflected light meets Target 1. Keep the mirror in this position.
5. Draw a ray by tracing a line on the paper along the straight edge of the mirror. Label the resulting angle with the corresponding target number.
6. Repeat the process for each target.

Day 2: Use the protractor to measure each angle. Record each measurement in degrees next to the corresponding angle. When students have finished measuring the angles, ask why the measurements varied. *(Students sat in slightly different locations; beams of light hit different spots on the paper targets.)*

Can You See Through It?

Materials:
flashlight for each small group
variety of transparent, translucent, and opaque items (such as writing paper, newspaper, plastic wrap, a tin can, plain glass edged with masking tape, frosted glass edged with masking tape, waxed paper, a magnifying glass, sunglasses, a plastic glove, construction paper, tissue paper, aluminum foil, a plastic cup, or a leaf)

To begin, review the definitions of the terms *transparent, translucent,* and *opaque* with students. Then divide students into small groups and give each group a flashlight. Have each group make a chart with headings like the one shown. Then have a child from each group choose an object from the selection of featured items. Direct the group to hold the item up to the flashlight and observe the amount of light that passes through it. Then have students list the item in the appropriate column of their chart. Direct each group to return the item to the designated location and then choose a different item. Have each group repeat this process with several different items.

Transparent (can see through clearly)	**Translucent** (can see through, but not clearly)	**Opaque** (cannot see through at all)
glass plastic wrap	waxed paper latex glove	aluminum foil white plastic lid

Transparent—a material that does not mix light rays and allows a person to see through it clearly

Translucent—a material that mixes light rays and allows a person to see through it, but not clearly

Opaque—a material that blocks all light rays, preventing a person from seeing through it at all

Curved Reflections

Materials:
variety of reflective items that are not flat planes, placed in a central location (such as a spoon, Christmas tree ball ornament, pan lid, bike reflector, shiny doorknob, makeup mirror, or sunglasses)

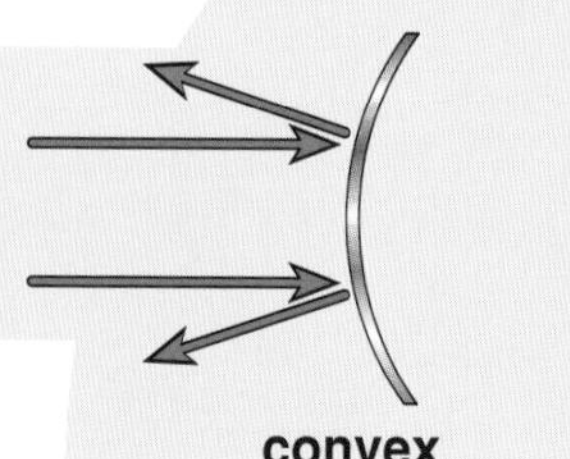

convex mirror

concave mirror

Challenge students to find out whether a curved surface reflects light in the same way as a plane surface with this hands-on activity. Discuss with students the difference between convex and concave mirrors. Point out that a convex mirror curves outward and makes objects appear smaller and farther away while a concave mirror curves inward and makes objects appear larger and closer. Then divide students into small groups and have each group make a chart with headings like the one shown. Have a child from each group choose one of the featured items. Guide the students in the group to test the item by holding it close to their faces and then at arm's length. Have the group record the results and classify the item as concave or convex. Then have each group return the item to the designated location and choose a different object. Direct each group to repeat this process with several different items.

Test Item	Description of Reflected Image	Classification
Inside of spoon	large, upside down	concave
Outside of spoon	small	convex
Shiny doorknob	small	convex
Makeup mirror	large	concave

Temperature-Raising Reflectors

This demonstration helps students see the effect concave mirrors can have on raising water temperature.

Materials:
two 12" x 18" sheets of poster board, one covered with aluminum foil
three 9 oz. plastic cups, each filled with water
3 thermometers
stacks of books

Steps:

1. Measure the water temperature in each cup and record it.
2. Place the cups of water in a sunny location outside.
3. Curve the foil-covered poster board halfway around one cup so that it faces the sun. Stack books behind the poster board to hold it in place.
4. Curve the plain poster board behind the second cup in the same way.
5. Place the third cup in a place where it will not be affected by either sheet of poster board.
6. Put a thermometer in each cup. Have students predict what will happen to the temperature of the water in each cup.

Follow-up: After 30 minutes, measure and record the temperature on each cup's thermometer. Discuss the results with students and compare them with students' predictions. *(The water temperature in the cup partly surrounded by the foil-covered poster board has the highest temperature because the shiny concave surface focuses the incoming light and heat onto the cup and its contents. The cup with nothing surrounding it has the coolest temperature.)*

High and Low Sounds

Explain to students that pitch is the degree of highness or lowness of a sound. Then guide student pairs through the steps shown to complete an experiment that explores pitch.

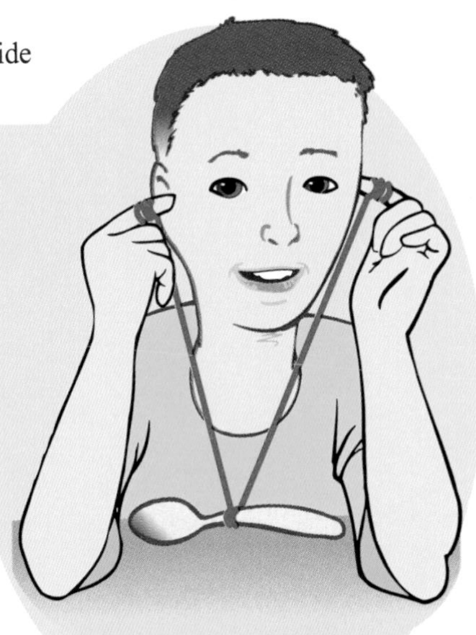

Materials for each pair of students:
metal spoon
wooden spoon
4' length of string

Steps:

1. Have Student 1 tie the spoon in the middle of the string as shown. Then have him wind the ends of the string around each of his pointer fingers, making sure the string is even on both sides of the spoon.
2. Direct Student 1 to carefully place a pointer finger in each ear. Have Student 2 gently strike the metal spoon with the wooden spoon.
3. Have Student 1 shorten the string by winding it around his fingers several more times. Then have Student 2 strike the metal spoon with the wooden spoon again.
4. Have the students switch roles and complete the experiment again.
5. Discuss how the two sounds compared. *(Students should respond that the pitch was lower when the string was longer and higher when the string was shorter. Pitch is determined by the number of vibrations made by a vibrating object in one second. The longer string caused a lower pitch because fewer vibrations were able to reach the ear in one second. The shorter string caused a higher pitch because more vibrations reached the ear in one second.)*

Bouncing Sound

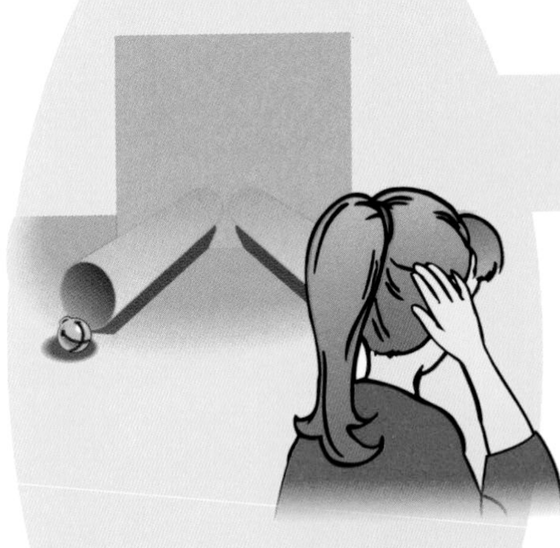

Divide students into groups of three and lead them through this hands-on experiment to explore the possibility that sound bounces.

Materials for every three students:
2 empty wrapping paper tubes
small bell
12" x 12" piece of cardboard
small pillow

Steps:

1. Have one student hold the cardboard vertically on a table while the other students place the tubes in front of it to form a V-shape.
2. Direct a student to sit with her back to the tube on the left and her ear against the tube on the right. Have the student cover her right ear with her hand.
3. Ask one student to jingle the bell into the tube on the left. Have the child with her ear to the tube on the right raise her hand when she hears the bell.
4. Have the group exchange the cardboard for the pillow. Then instruct the group to repeat Steps 2 and 3.
5. Have the group members trade places until each student has had a chance to listen to the bell reflected off both objects.
6. Discuss the groups' observations as a class. *(Students should conclude that the sound of the bell bounced, or reflected, off the cardboard and the pillow, but the cardboard reflected the sound better. A hard, smooth surface like cardboard reflects sound better than an uneven, soft surface like a pillow. This is because the soft surface absorbs much of the sound.)*

Listen to This

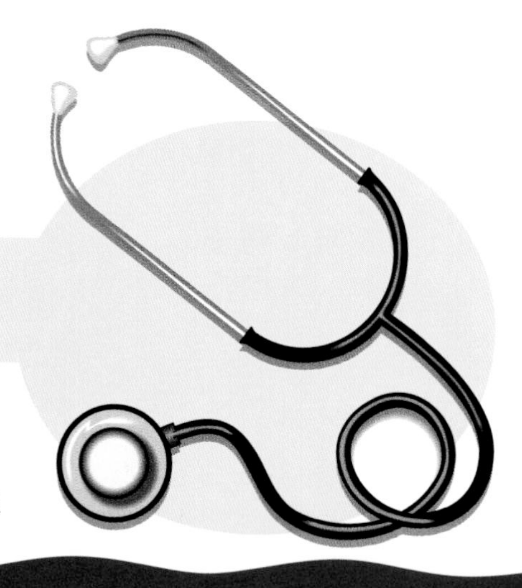

Materials for each student:
copy of page 55
piece of poster board

Help students gain a better understanding of amplification with this hands-on experiment. Explain to students that throughout history, devices have been created to help capture sound waves and direct them closer together, making them easier to hear. Then pair students and have them follow the steps on page 55 to make a megaphone and complete the experiment.

Did You Know?
In 1819, a French physician invented the stethoscope to listen to a patient's heart. It helped amplify (make louder) quiet sounds in the body.

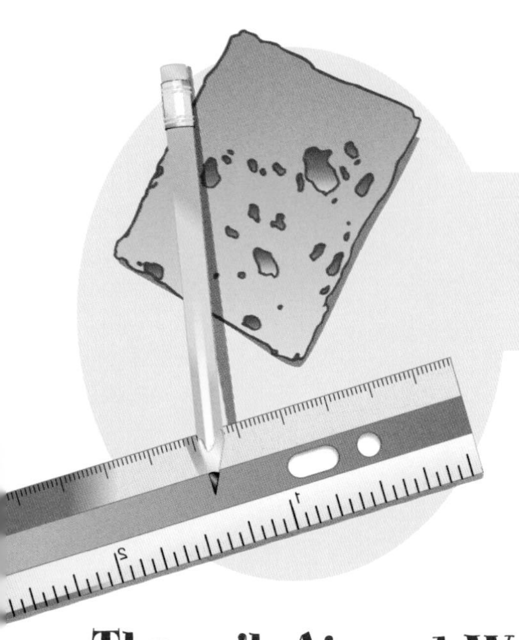

Sound Through Solids

Materials:
copy of page 56 for each student
pencil for each pair of students
variety of objects made from different materials, placed in a central location (such as a plastic ruler, wooden ruler, sponge, jacket, book, or cleaned and sanitized egg carton)

Help students find out whether sound can travel through solids. Explain to students that conductors are objects that allow sound to travel through them. Then give each child a copy of page 56. Pair students and have them follow the steps on the page to complete the experiment. When each duo has completed the experiment, lead students in discussing the results. *(Students should conclude that they were able to hear the sounds through the table better than through the air. Explain that this is because solid materials, such as wooden desks, are good conductors of sound.)*

Through Air and Water

Materials for each pair of students:
2 resealable plastic bags: one ¾ full of water and one full of air

Remind students that a conductor is a substance that allows sound to travel through it. Have one child in each duo hold the water-filled bag to his ear and tap the bag lightly. Then direct him to do the same thing with the air-filled bag. Ask each student if he can hear the tapping through the bags. *(Students should respond that they could hear the tapping. Explain that this is because water and air are both good conductors of sound.)*

Did You Know?
Whales keep in contact with each other by singing songs, which travel hundreds of miles through the sea.

See the light and sound skill sheets on pages 57–60.

Name ______________________________

Listen to This!

Follow the directions below to make a simple megaphone.

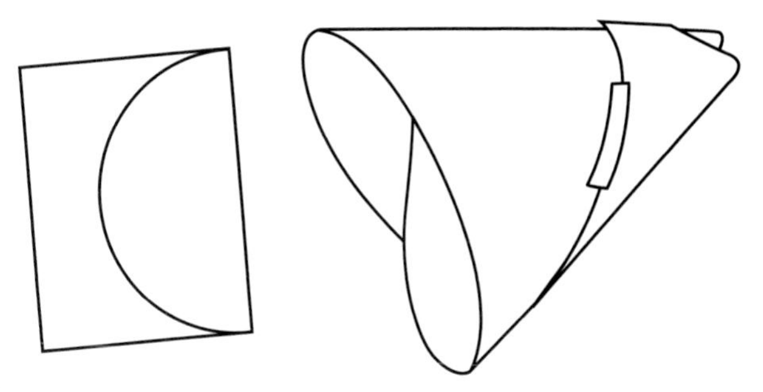

1. Draw a half circle on a piece of poster board, cut it out, roll it into a cone, and tape it in place.
2. Record your hypothesis, or what you think will happen when you put the cone to your ear and listen. Then record what you think will happen when you put the cone to your mouth and speak through it. ______________________________

   ____

   ____

   ____
3. Ask your partner to stand about six feet away and speak to you in a normal voice. Listen to the loudness of his or her voice.
4. Place the small end of the cone carefully up to your ear and listen again. Is there a difference in loudness? ______________
5. Record your observations, or what you heard with and without the cone up to your ear. Did the cone amplify the sound of your partner's voice? ______________________________

   ____
6. Speak to your partner in a regular voice. Place the small end of the cone up to your mouth and repeat what you said. Did your partner notice a difference in the loudness of your voice?

   ____
7. Record what your partner heard with and without the cone up to your mouth. Did the cone amplify the sound of your voice? ______________________________

   ____

   ____

   ____
8. Trade places with your partner and repeat the experiment.

Note to the teacher: Use with "Listen to This" on page 54.

Name________________________

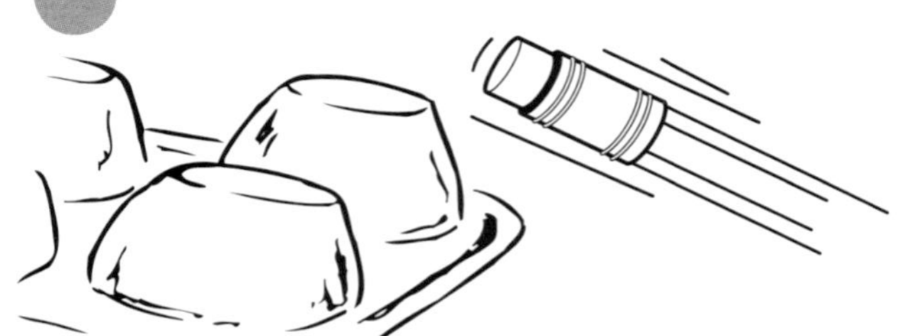

Sound Through Solids

Some materials allow sound to travel through them easily. These are called *conductors.* Follow the steps below to learn more about solids as sound conductors.

A. 1. Put your head down on the desk so that one ear is on the desk. Cover your other ear with your hand.
2. Have your partner place an object on the desk and record its name on the chart. Then have your partner use a pencil to tap the item.
3. Listen carefully. How well does the sound travel through the desk?
4. Lift your head and listen as your partner taps the object once again. How well does the sound travel through the air?
5. On the chart, rate the loudness of each of the sounds.
6. Repeat Steps 1–5 with different objects.
7. Trade places with your partner and repeat Steps 1–5.

B. On the back of this sheet, explain why you think you got the ratings you did on the chart. Keep in mind what you have learned about the way sound travels.

Object	Sound Through Solid	Sound Through Air
	not at all — very well 0 1 2 3 4 5	not at all — very well 0 1 2 3 4 5
	not at all — very well 0 1 2 3 4 5	not at all — very well 0 1 2 3 4 5
	not at all — very well 0 1 2 3 4 5	not at all — very well 0 1 2 3 4 5
	not at all — very well 0 1 2 3 4 5	not at all — very well 0 1 2 3 4 5
	not at all — very well 0 1 2 3 4 5	not at all — very well 0 1 2 3 4 5
	not at all — very well 0 1 2 3 4 5	not at all — very well 0 1 2 3 4 5
	not at all — very well 0 1 2 3 4 5	not at all — very well 0 1 2 3 4 5

Note to the teacher: Use with “Sound Through Solids” on page 54.

Name ______________________________

What Are They Looking At?

Each child below sees a different object in the mirror. Mirrors reflect light. The angle at which light strikes a mirror is called the angle of incidence. The angle at which light reflects from the mirror is called the angle of reflection. Follow the directions below to match each child to the object he or she sees. If you match them correctly, you'll learn another name for a flat mirror!

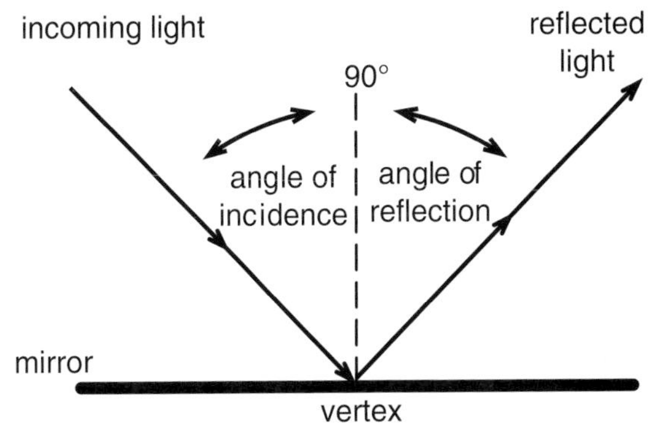

1. Place a protractor on the dotted line below. Align the center of the protractor with the vertex, the point where all the angles hit the mirror. Measure the angle between the cat and the dotted line. Record the measurement in degrees in the blank next to the cat.
2. Keep the protractor in place and measure and record the angles of the cake, apple, banana, and dog.
3. Using the straight edge of the protractor, draw line segments connecting the vertex on the mirror with the nose of each child.
4. Repeat Step 1, measuring the angle between the dotted line and each child. Record each measurement in degrees in the top blank next to each child.
5. Identify the object each child is looking at by matching the number of degrees. Then write the name of the object in the second blank.

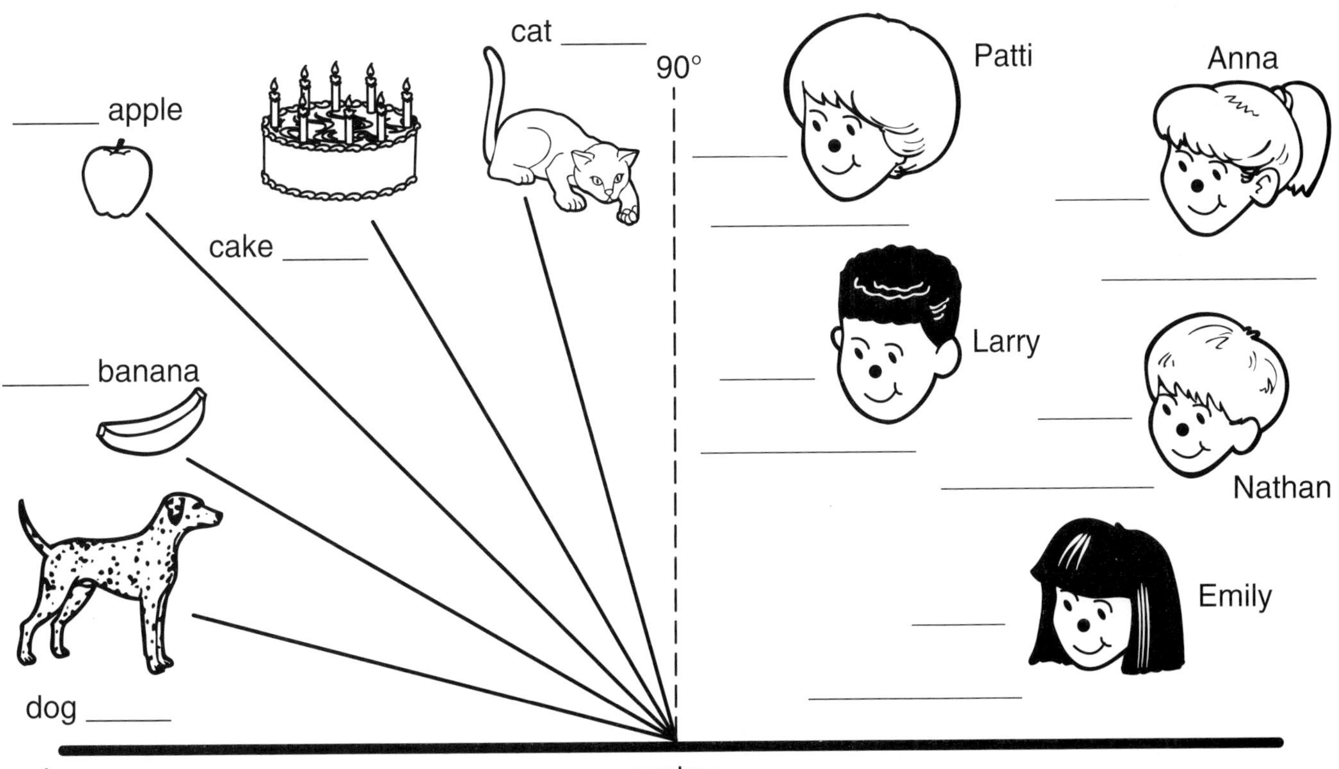

Now learn another name for a flat mirror by writing the first letter of each child's name in the following blanks in order from least to greatest number of degrees.

___ ___ ___ ___ ___

Name ______________________________ Investigating mirror images

Flip-Flopped Messages

Does the writing on the mirrors below look strange? That's because images in mirrors are reversed. Find out what each flip-flopped message says by placing a mirror on the corresponding dotted line. Record each decoded message on the back of this page. Then do what each message directs you to do!

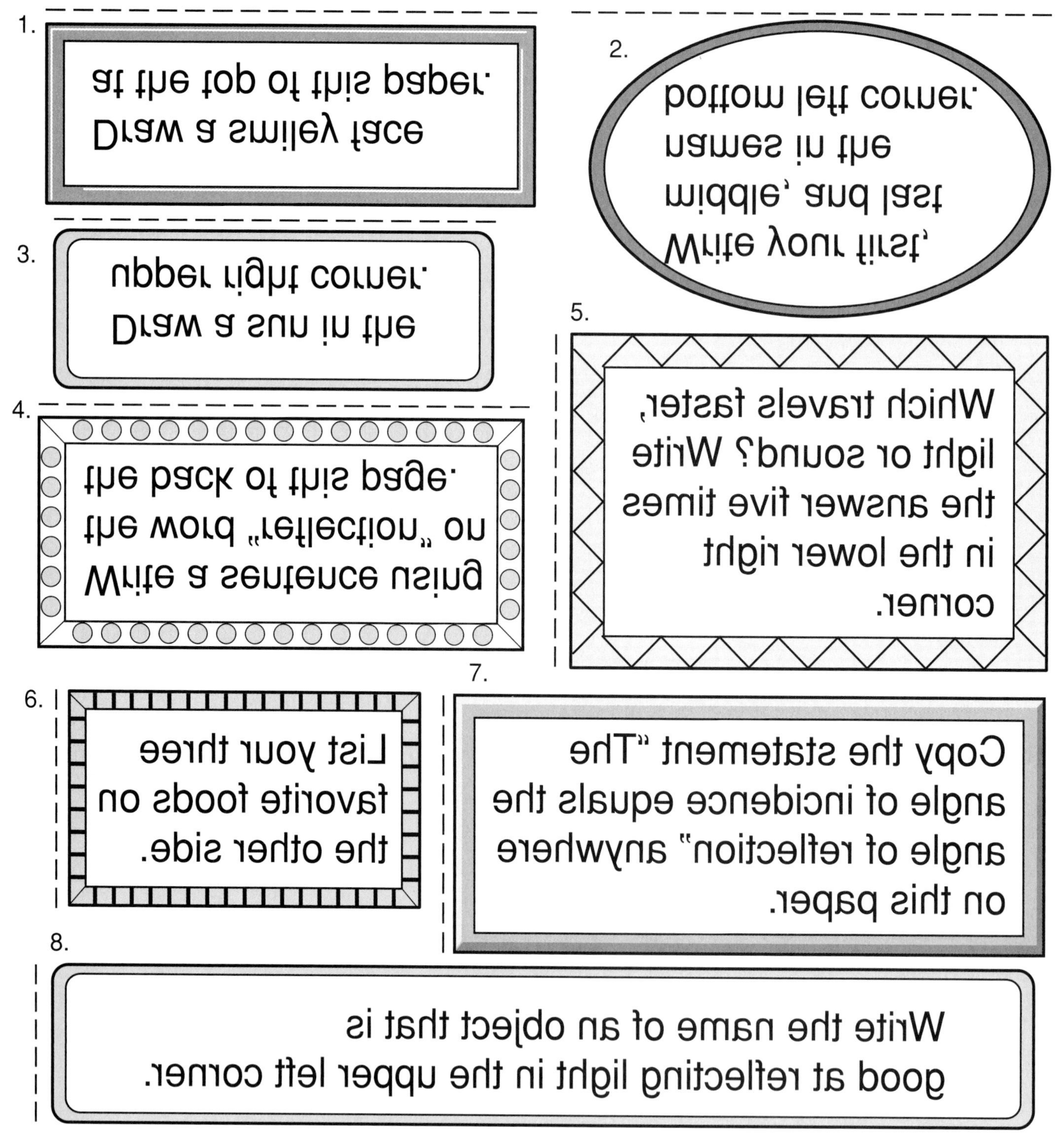

Bonus: Leonardo da Vinci, an Italian scientist and artist, used mirror writing to record many of his notes. Write a message on a sheet of paper. Next, place a mirror on the paper above the message and copy its reflected words onto another sheet of paper. Then challenge a classmate to use a mirror to decode the message.

Name____

Detecting Decibels

Detective Dessie Bell has lost all of the sounds from the chart below. Use your detective skills to help Dessie Bell put her sounds back in order!

Directions: The chart below contains nine decibel levels, but the name of a sound made at each level is missing. Read each clue below to figure out where each sound belongs. Then write the name of the sound in the corresponding space on the chart. **Remember:** A *decibel (dB)* is used to measure the intensity level of a sound.

Decibel Levels

140 ________________________________

120 ________________________________

100 ________________________________

90 ________________________________

80 ________________________________

70 ________________________________

60 ________________________________

20 ________________________________

0 ________________________________

Clues:

1. The threshold of audibility is the level of the weakest sound that can be heard by the human ear.
2. The threshold of pain is the level of the loudest sound that can be heard and may cause pain in the ear.
3. A telephone ring is more intense than a conversation, but not as intense as a vacuum cleaner.
4. Heavy traffic is more intense than a vacuum cleaner, but not as intense as a circular saw.
5. An amplified rock concert is more intense than a circular saw, but not as intense as the sound of a jet takeoff at close range, which is at the same level as the threshold of pain.
6. A whisper is not as intense as a conversation.

Bonus: Think about the numbers that are missing from the chart, such as 10, 30, and 40. On the back of this sheet, make a new chart, including the missing numbers. Then fill in the rest of the chart with sounds you think might fit at each level.

Name______________________________

I Have Your Frequency

Frequency, measured in hertz, is the number of vibrations made by a vibrating object per second. Humans can make sound frequencies from 85 to 1,100 hertz. Some animals can emit higher frequencies than humans, and some can emit lower frequencies than humans. Study the frequency chart below and then answer the questions that follow.

Frequency Chart	
Animal	**Hertz Range**
Cat	760–1,520
Grasshopper	7,000–100,000
Bat	10,000–120,000
Dog	452–1,080
Dolphin	7,000–120,000
Human	85–1,100
Robin	2,000–13,000

1. Which animal has the lowest frequency range? ______________________
   Which two have the highest? ______________________
   (Hint: Look at the first number for lowest and the last number for highest.)
2. Which animal has the widest frequency range? ______________________
   (Hint: Subtract the lowest number from the highest number for each animal.)
3. Which animal has the narrowest frequency range? ______________________
   (Hint: Solve the same way as in number 2.)
4. How much wider is a bat's frequency range than a human's frequency range (in hertz)?
   ____
5. How much wider is a cat's frequency range than a dog's frequency range (in hertz)?
   ____
6. Name the animals that have a wider frequency range than a human has. ______________________
   ____

Bonus: On the back of this sheet, list the animals in order from widest frequency range to narrowest frequency range.

Forms of Energy

What a Stretch!

Materials for each group of students:
copy of the top half of page 66
3 same-size rubber bands
ruler
yardstick

Help students understand potential and kinetic energy with rubber bands! In advance, make a masking tape starting line on the classroom floor. Then explain to students that kinetic energy is the energy of motion and potential energy is energy that is stored. Demonstrate to students how to stretch the rubber band as shown. Tell students that the stretched rubber band is an example of potential energy, and that releasing the rubber band is an example of kinetic energy. Before students begin the experiment, remind them to aim their rulers straight ahead and to avoid releasing a rubber band if a classmate is in its path. Direct students to complete the experiment on page 66. After the experiment, lead students in discussing how the stretch of the rubber band affects the distance it travels. *(As you stretch the rubber band, you increase its potential energy. This results in a greater kinetic energy, which makes the rubber band travel faster when released.)*

Swing Time

Materials for each pair of students:
copy of the bottom half of page 66
5 metal washers
16" length of string
unsharpened pencil
jumbo paper clip
masking tape

This idea not only demonstrates potential and kinetic energy, but also explores the science behind a pendulum! Demonstrate the steps on page 66, pointing out to students that the pencil, string, and paper clip make a pendulum. Then have each pair follow the steps on page 66 to complete the experiment. When students are finished, explain that the potential energy in the experiment is stored in the paper clip and washers. When the paper clip and washers are released, their potential energy converts into kinetic energy. Discuss the results of the experiment and lead students to understand that the number of swings stays basically constant regardless of the number of washers used. Explain that this happens because a pendulum will take the same amount of time to make every swing regardless of how heavy the weight at the end of it is.

Did You Know?
In a grandfather clock, a pendulum of 39 inches will swing back and forth 60 times in one minute, resulting in an accurate measure of time.

Quick-Fade Quilt

Invite students to witness the power of light energy with this sunny investigation.

Materials for each student:
4" construction paper square
4" tagboard square
paperweight (such as a small rock)

Have each student cut a sun shape from his tagboard square and set his cutout atop his construction paper. Direct each student to set the project outside in a sunny spot and place a paperweight atop the sun cutout. Leave the papers in place for one hour or more.

Follow-up: At the end of the time period, have each student retrieve his project from outside and remove his sun cutout; then discuss the results with students. *(The paper hidden by the cutout is darker, indicating that the exposed construction paper has faded.)* Point out that the light energy from the sun caused a chemical reaction in the paper. If desired, have students glue their construction paper squares on a large sheet of bulletin board paper to make a giant class quilt.

Hot, Hot, Hot!

Materials:
small bowl of water
thermometer
¼ c. hydrogen peroxide
1 tbsp. quick-rising dry yeast
spoon

Demonstrate a chemical change that produces energy with this activity. Record the temperature from the thermometer on the board. Then place the thermometer in the bowl. Pour the hydrogen peroxide and yeast into the bowl. Stir the mixture and have students observe what happens. *(The solution bubbles up.)* Have a volunteer feel the sides and bottom of the bowl and tell the class how they feel. *(They should feel warm.)* After two minutes, use the spoon to remove the thermometer from the bowl. Record the thermometer's current reading on the board. Ask students why they think the temperature rose. *(A chemical change that produced heat took place.)* Explain that when the yeast was added, the hydrogen peroxide changed into oxygen and water molecules. The bubbles were created by oxygen gas escaping during the change. The chemical change also produced heat, proven by the warmth of the bowl and the thermometer reading

Food Fuel

Materials for each student:
copy of page 67

A great example of chemical energy is in the human body. Explain to students that food they eat combines with oxygen, causing a chemical change that releases heat and energy. This heat energy is measured in calories. Tell students that the number of calories listed on a food item's container tells how much energy the food provides when it is completely used by the body. Then direct each child to complete her copy of page 67.

Person: Max
Age: 4

Meal	Calories
Breakfast:	
apple juice (1 c.)	116
small bagel	64
	180
Lunch:	
turkey sandwich	400
orange	62

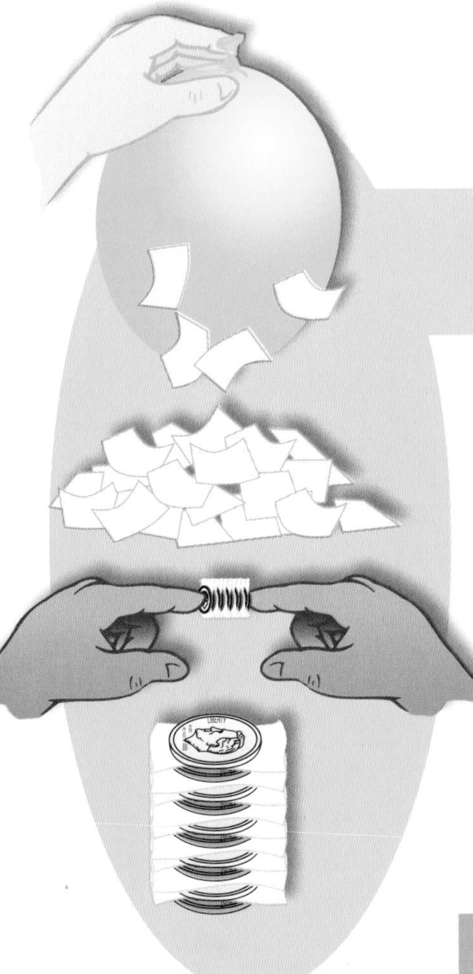

Making Connections

Help students distinguish between static electricity and current electricity with the following demonstrations.

Materials:
piece of nylon
inflated balloon
ripped-up piece of paper
nine 1" x 2" strips of paper towel soaked in lemon juice
5 dimes
5 pennies

- **Charge It!** (static electricity) Rub the nylon against the balloon; then bring the balloon near the head of a student with long hair. Have students observe what happens. *(The hair will be attracted to the balloon).* Next, bring the balloon near the pile of paper pieces. Have students observe what happens. *(The paper will be attracted to the balloon.)* Ask students what they think caused these results. Then explain that when two different substances are rubbed together, electrons may jump from one material to the other. This type of electricity doesn't flow and is called static electricity.

- **How Shocking!** *(current electricity)* Stack the coins, alternating the dimes and pennies, placing a juice-soaked strip between each coin. Invite a volunteer to moisten one fingertip on each hand in water and hold the coins between her fingers as shown. Have the student describe what she feels. *(She should feel a small shock or tingle.)* Explain that the lemon juice, an acid, conducts the electricity created by the coins' metals. The stack of coins acts as a wet cell, an early form of the batteries sold in today's stores. A real battery is made up of two or more dry cells, each containing metals separated by paper soaked in a strong acid.

Did You Know?
Electricity is energy that is created by the flow of electrons. Static electricity is the buildup of positive or negative charges on an object. Current electricity is the flow of electrons in a conductor through a closed circuit.

Hot or Cold?

This demonstration shows students how an object's size can be affected by a temperature change.

Materials:
plastic water bottle
2 reusable, deep plastic containers, one filled with 2–3 c. of heated water and one filled with 2–3 c. of ice water
balloon

Steps:
1. Place the neck of the balloon over the opening of the water bottle.
2. Set the bottle in the hot water and observe the balloon for one minute. *(The balloon should begin to inflate, or expand.)*
3. Set the bottle in ice water and observe the balloon for one minute. *(The balloon should begin to deflate, or contract.)*

Conduction Instruction

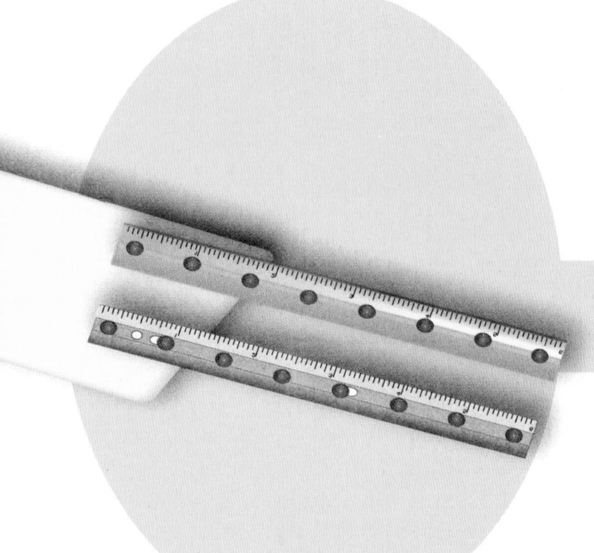

Use this demonstration to show students which materials are best at allowing heat to travel through them. Begin by explaining to students that heat can travel through a material by conduction. Tell students that not all materials are good conductors of heat. For example, when a metal pot is heated on a burner, the metal becomes hot, but the plastic handle stays cool. This is because metal is a better conductor of heat than plastic. Follow the steps below to demonstrate which material is a better conductor of heat.

Materials:
wooden ruler
metal ruler
heating pad
chocolate chips

Steps:
1. Set one end of each ruler on the heating pad.
2. Place eight chocolate chips on each ruler, evenly spaced apart.
3. Turn the heating pad to high.
4. After 15 minutes, have students observe which ruler has melted more chocolate chips. *(The metal ruler should melt more chocolate chips because it conducts heat better than the wooden ruler.)*

Did You Know?
During conduction, heat causes atoms in a material to quickly vibrate and strike each other. As they meet, these atoms transfer heat from one to another, spreading the heat through the materials.

Spinning Spirals

Explain to students that heat can pass from one object or place to another by *convection.* For example, when the heat from a lamp, stove, or heater moves and rises and then cools and sinks, a convection current is created. Guide students through the steps shown to observe a convection current.

Materials:
class supply of copies of page 68
length of string for each child
lamp with shade removed, placed on the floor and turned on in advance to heat the bulb

Steps:

1. Have each child decorate his spiral and then cut it out along the bold line.
2. Direct each child to use the tip of a sharpened pencil to poke a small hole in the top of the spiral where indicated. Then have him thread the string through the hole and knot it.
3. Have each student, in turn, hold his spiral above the hot lamp, making sure not to touch the paper to the bulb.
4. When the spiral begins spinning, explain that the air above the bulb is heated, and a convection current is created.

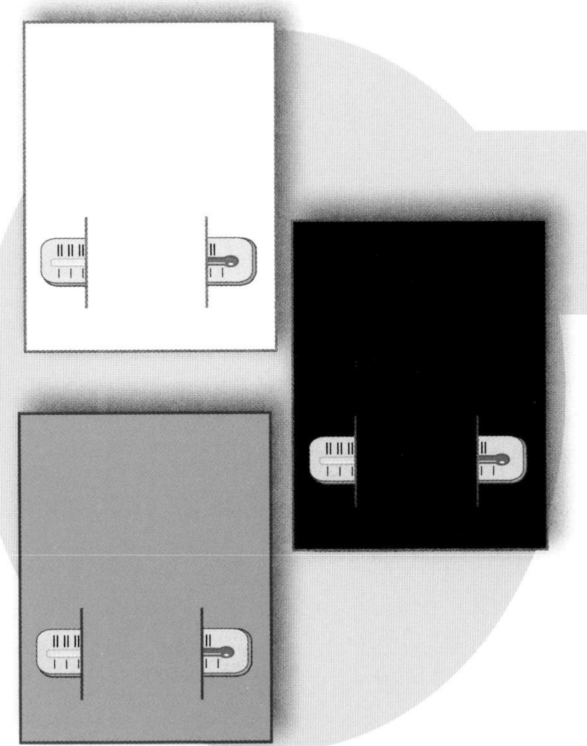

Hot Colors

Materials:
three 9" x 12" sheets of construction paper (one black, one white, and one green) with slits cut on each as shown
3 thermometers
access to a warm, sunny location

Show your students the importance of color in capturing solar energy with this experiment. Place the papers in a sunny location and slide a thermometer into each slit. Record the starting temperature of each thermometer. After about 10 minutes, record the temperature on each thermometer again and lead students in discussing the results. *(The thermometer in the black paper should show the highest temperature, while the thermometer in the white paper should show the lowest temperature.)* Explain to students that dark colors like black absorb most of the sun's heat, while light colors like white reflect most of the sun's heat. Other colors like green only absorb a small part of the sun's heat. Lead students to understand that this is why most items that use solar energy, like solar water heaters, are black.

Name(s) ____________________ Exploring potential and kinetic energy

What a Stretch!

Directions:

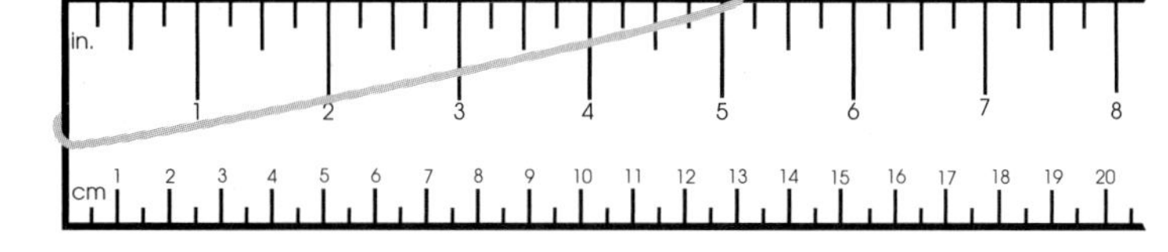

1. Stretch a rubber band around one end of the ruler as shown. Pull it back to the five-inch mark.
2. Stand behind the starting line. Release the rubber band. Measure the distance the rubber band traveled in inches and record the data below.
3. Repeat Steps 1 and 2 for a total of three trials.
4. Repeat Steps 1–3, stretching the rubber band back to the six- and seven-inch marks.

	Distance Traveled in Inches			
Rubber Band Stretch	Trial 1	Trial 2	Trial 3	Mean
five inches				
six inches				
seven inches				

Conclusion: How did the stretch of the rubber band affect the distance it traveled? ____________________

____

____

Name(s) ____________________ Demonstrating potential and kinetic energy

Directions:

1. Tape the pencil to a desktop so that half of it extends off the desk.
2. Tie one end of the string to the extended end of the pencil. Tie the other end to the paper clip. Slip one washer onto the paper clip.
3. Pull the paper clip so that the string is parallel with the desktop. Then release it. Partner 1 keeps time while Partner 2 counts the number of swings in ten seconds (forward and back = one swing).
4. Add a second washer to the paper clip. Repeat Step 3.
5. Continue until you have completed three trials each with one, two, and three washers.
6. Calculate the mean number of swings for each number of washers.

	Number of Swings in Ten Seconds			
Washers	Trial 1	Trial 2	Trial 3	Mean
1				
2				
3				

Note to the teacher: Use "What a Stretch!" with the first idea on page 61. Use "Swing Time" with the second idea on page 61.

Name ______________________________

Counting Calories

The movement of our bodies causes the food we eat to chemically change into energy. Different foods contain different amounts of energy. This energy is measured in units called *calories.*

Below are recommendations for how many calories a person needs each day. Also included is a list of foods and the number of calories in one serving of each. Study the list. Then follow the directions below.

Approximate Recommended Daily Calorie Needs

Children			Males	Females
1–3 years	1,000	11–14	2,200	2,000
4–6	1,300	15–18	2,600	2,000
7–10	1,700	19–22	2,700	2,100
		23–50	2,500	2,000

Food	Calories	Food	Calories	Food	Calories
apple juice (1 c.)	116	corn on cob	125	pretzel sticks (40)	100
small bagel	64	egg	77	rice (1 c. white)	216
banana	109	french fries (reg.)	237	sirloin steak (3 oz.)	177
broccoli ($\frac{1}{2}$ c.)	23	grape juice (1 c.)	120	spaghetti (1 c. noodles, $\frac{1}{2}$ c. tomato sauce)	238
brownie (2 oz.)	243	ice cream (1 c. chocolate)	300	strawberries (1 c.)	45
carrots (2 med.)	62	mac & cheese (1 c.)	430	toast (1 slice white bread)	67
cheeseburger	318	milk (1 c. skim)	86	turkey sandwich	400
cheese pizza (2 slices)	492	onion rings (8)	175	waffle	108
chicken (3 oz.)	150	orange	62	yogurt (1 c.)	225
chocolate cake	210	potato chips (1 oz.)	140		
corn flakes (1 c.)	88				

Directions: On the back of this page or another sheet of paper, use the calorie recommendations and food chart to complete a menu like the sample shown for two of these people:

- Max, age 4
- Mindy, age 11
- Mr. Goodbody, age 32
- Mrs. Morse, age 49

Add to find the total calories for each menu. Make sure you don't go over the daily recommendation.

Sample Menu

Person: ______________

Age: ______________

Meal — Calories (list foods for each)

Breakfast: ______

Lunch: ______

Dinner: ______

TOTAL CALORIES = ______

Note to the teacher: Use with "Food Fuel" on page 63.

Spiral Pattern

Use with "Spinning Spirals" on page 65.

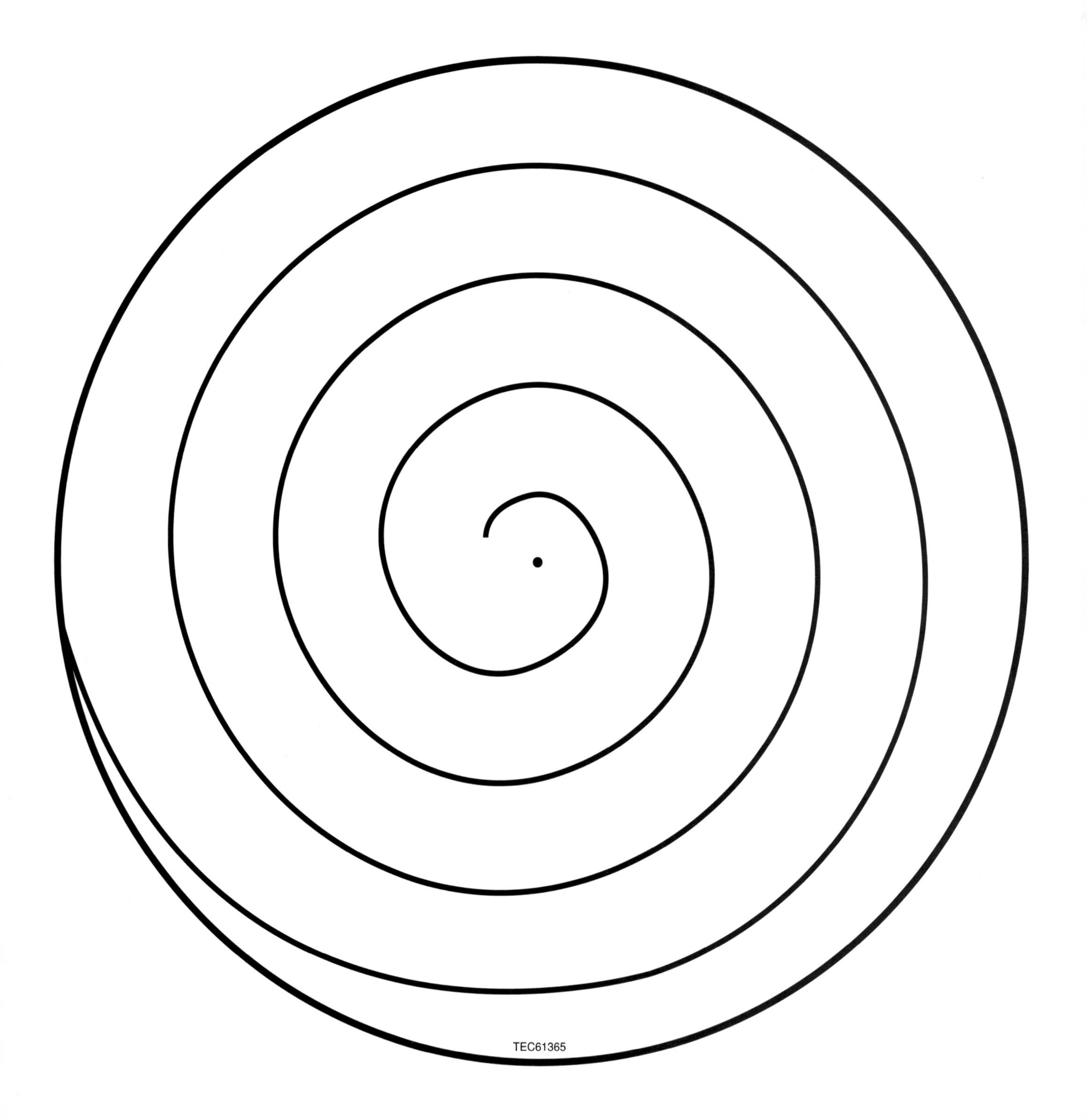

Laws of Motion

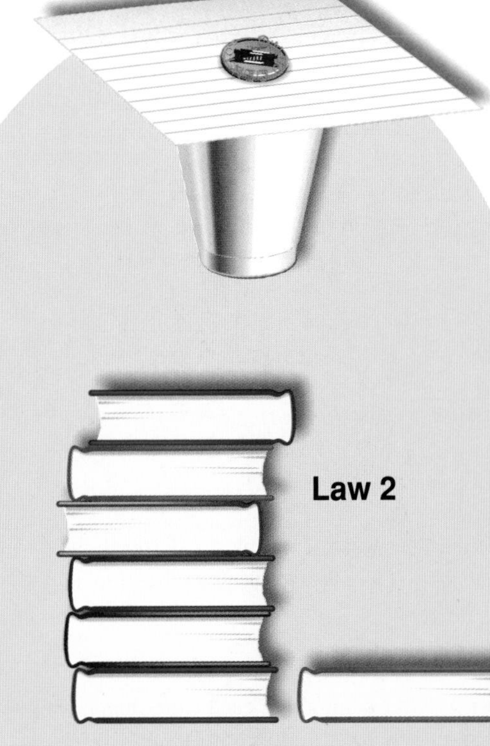

Help students understand Newton's laws of motion with these simple demonstrations. Begin each demonstration by telling students about the corresponding law. Conclude each demonstration by having students complete the corresponding section on page 72.

Materials:

student copies of page 72
penny
plastic cup
5 marbles
index card
ruler (with a groove)
7 textbooks

Day 1: Law 1—An object at rest will remain at rest and an object moving in a straight line will continue moving in a straight line unless acted upon by an outside force. Place an index card atop a cup and a penny in the center of the card. Have students observe what happens when the card is quickly pulled away. *(The card is removed from the cup, but the penny drops into the cup.)* Explain that gravity, the only force working on the penny, pulls the penny down into the cup.

Day 2: Law 2—The change in motion of an object depends on its mass and the force acting upon it. Stack six textbooks next to a single textbook. Ask students which stack is harder to push. Have a student volunteer push the books to show that six books are harder to move then one. Explain that the greater the mass, the greater the force needed to move it.

Day 3: Law 3—For every action, there is an equal and opposite reaction. Align the marbles in the ruler's groove. Move one marble away from the others and then roll it slowly toward the others to see what happens. *(One marble at the opposite end moves.)* Repeat the demonstration, this time rolling two marbles slowly toward the others, forcing two marbles at the opposite end to move. Explain that the reaction is always equal to the action, with the force of the roll being transferred through the row of marbles to those on the end (one marble moving one or two marbles moving two).

Fabulous Friction

Explain to students that friction is a force that tries to stop movement. Then conduct the following experiments to demonstrate how friction can be increased or decreased.

Experiment 1

Experiment 2

Experiment 3

Materials:

brick
12 marbles
metal jar lid
cardboard square
6 rounded (not hexagonal) pencils

Experiment 1: Place the brick on the cardboard. Ask students if they think the brick will move easily on the cardboard. Then have a student give the brick a small push to see what happens *(little, if any, movement)*. Ask students what makes the brick difficult to slide across the cardboard. *(The friction, which is caused by the brick's rough texture and weight, makes the brick difficult to slide across the cardboard.)*

Experiment 2: Place the pencils side by side on a flat surface and lay the brick atop them. Have a student give the brick a small push to see what happens. *(The brick moves easily.)* Ask students to explain why. *(The pencils work as rollers and reduce the friction because less of the brick's surface touches another surface. Smooth surfaces also produce less friction than rough surfaces.)*

Experiment 3: Place the marbles underneath the jar lid and the brick atop the lid. Spin the brick to let students see that the brick spins freely. Have them explain why. *(The marbles act as ball bearings that reduce the friction and allow the brick to spin easily.)*

Follow-up: After completing each experiment, lead students in discussing ways friction can be increased or decreased in different situations.

Sports Motions

Use this critical-thinking activity to help students relate Newton's laws of motion to movements in sports.

Day 1: Invite students to brainstorm different sports movements. Then choose one movement, such as dribbling a basketball. Explain to students that the force used to push the ball to the floor equals the force with which it bounces back. Further explain that the basketball's texture increases friction, making it easier to hold and control. Then divide students into groups and assign each group a different sport. Have each group discuss and list the movements associated with its sport.

Day 2: Have each group review its list from the previous day. Then direct them to list at least ten ways the laws of motion affect movement in their chosen sport, including the effects of gravity and friction. Finally, invite each group to present its observations to the class.

Moving Right Along

Demonstrate Newton's third law of motion with this balloon-powered vehicle. To begin, remind students that Newton's third law of motion states that for every action there is an equal and opposite reaction. Then follow the steps below to show how a balloon expels gas (the action) to propel itself forward in the opposite direction (the reaction).

Materials:
drinking straw
12' length of string
2 chairs
binder clip
balloon
tape

Steps:
1. Thread the string through the straw. Tie one end of the string to the back of a chair as shown. Tie the other end of the string to another chair. Move the chairs apart so the string is taut.
2. Inflate the balloon and close it with a binder clip.
3. Tape the balloon to the straw as shown.
4. Gently pull the straw until the clipped end of the balloon almost touches the chair. Release the binder clip. Repeat as desired.

Vocabulary in Motion

Materials for each student:
list of motion words for them to define (see suggestions)

Here's a fun way to help students learn the meanings of motion-related words. To begin, give each student the list of words and direct her to write the definition for each word. Then pair students and secretly assign each duo two to three words. Have the partners work together to plan a skit in which they act out each word's meaning (without saying any form of their assigned words) for the class to guess.

Motion Words

force
motion
friction
inertia
momentum
gravity
kinetic energy
potential energy
resistance
mass
center of gravity
perpetual motion

Name ______________________

Isaac's Incredible Laws

Illustrate an experiment for each law. Then, write at least two sentences explaining how the experiment illustrates that law.

Isaac Newton's Laws of Motion

Law 1: Unless acted upon by an outside force, an object at rest will remain at rest and an object moving in a straight line will continue moving in a straight line.

Law 2: The change in an object's motion depends on its mass and the force acting upon it.

Law 3: For every action, there is an equal and opposite reaction.

Bonus: Find a magazine picture that illustrates each of Newton's laws of motion. Glue the pictures to a sheet of paper. Add a caption to each picture explaining how it illustrates that law.

Note to the teacher: Use with "Laws of Motion" on page 69.

Electric Activities

Materials for each student:
sheet of unlined paper

Have students create electricity usage charts to explore the many ways they use electricity. First, have students brainstorm different uses of electricity at home, at school, and elsewhere as you list their ideas on the board. Next, direct each child to create an electricity-usage chart by dividing a sheet of paper into four columns and labeling the headings as shown. Instruct each child to list her activities on a typical day and then complete the remaining chart columns. When students are finished with their charts, discuss the alternatives they came up with to replace activities that require electricity.

Activity	Electrical appliance used	Can the activity be completed without electricity?	If yes, how? If no, why?
Waking for school on time	Alarm clock	Yes	Use a wind-up clock.
Watching cartoons	Television	No	Television requires electricity to operate.

Simple Circuit

Lead students in this whole-group activity to demonstrate the flow of electricity. Explain to students that a circuit is a complete path in which an electrical current flows. Have students stand in a circle and hold hands. Tell students that you will squeeze the student's hand on your right to represent the flow of electricity. Then direct each student to squeeze the student's hand to his right as soon as he feels the squeeze from the person on his left. Explain that this represents the flow of electricity through a closed circuit, or electricity (squeezing hands) flowing through a wire (students holding hands) from a power source (teacher) and back to complete the circuit.

Repeat the demonstration, but remove yourself from the circle before the "electricity" reaches you. Ask students to describe what happens when the circle is broken. *(The hand squeezing comes to a stop at the break in the circle.)* Explain to students that electric current only flows when it can follow a closed path, or circuit. Ask students to brainstorm examples of everyday items that have closed circuits, such as hair dryers, lamps, and televisions. List students' responses on chart paper titled as shown.

Electrical Circuits All Around Us

DVD player
toaster
lamp
television
coffee maker
iron
sound machine
mixer
hair dryer
straightening iron

Kilowatt Hours

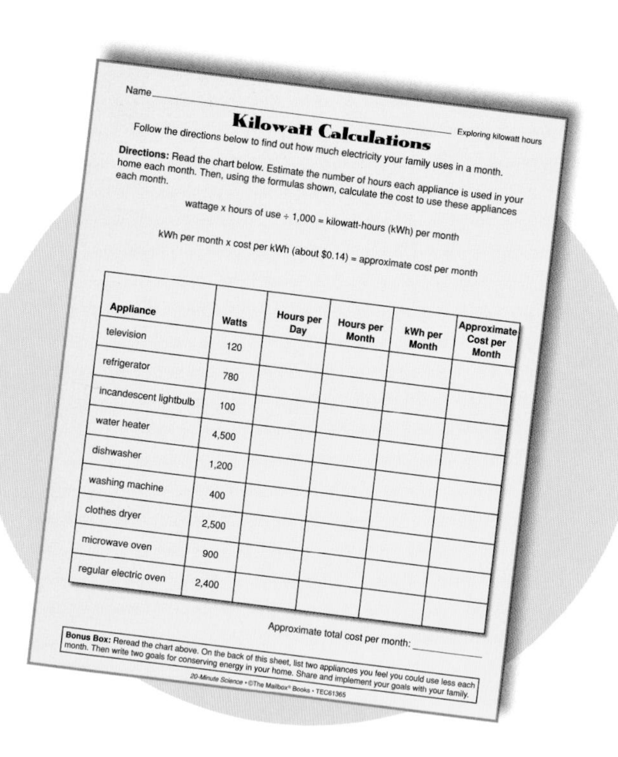
Name ______

Kilowatt Calculations

Exploring kilowatt hours

Follow the directions below to find out how much electricity your family uses in a month.

Directions: Read the chart below. Estimate the number of hours each appliance is used in your home each month. Then, using the formulas shown, calculate the cost to use these appliances each month.

wattage x hours of use ÷ 1,000 = kilowatt-hours (kWh) per month

kWh per month x cost per kWh (about $0.14) = approximate cost per month

Appliance	Watts	Hours per Day	Hours per Month	kWh per Month	Approximate Cost per Month
television	120				
refrigerator	780				
incandescent lightbulb	100				
water heater	4,500				
dishwasher	1,200				
washing machine	400				
clothes dryer	2,500				
microwave oven	900				
regular electric oven	2,400				

Approximate total cost per month: ______

Bonus Box: Reread the chart above. On the back of this sheet, list two appliances you feel you could use less each month. Then write two goals for conserving energy in your home. Share and implement your goals with your family.

Materials for each student:
copy of page 75

Help students learn about kilowatts with this reproducible activity. Explain to students that electricity is measured in kilowatts, or the unit of measure equal to 1,000 watts. Further explain that a kilowatt-hour is the unit of energy expended by one kilowatt in one hour. Ask students how much they think a kilowatt hour of electricity costs (about $0.14). Then give each child a copy of page 75 to complete as directed.

Swinging Magnets

This experiment shows students how magnetic poles attract and repel each other. After each pair conducts the experiment, invite them to share their conclusions. *(Students should conclude that opposite poles—north and south—attract each other and like poles—north to north and south to south—repel each other.)*

Materials for each pair of students:
2 bar magnets with the poles marked
length of string
tape
paper

Steps for students:

1. Read Steps 2–7 and then record your hypothesis, or what you think will happen as you move the magnets toward each other. Then complete Steps 2–8.
2. Tie the string around the ends of the magnet as shown.
3. Tape the middle of the string to the edge of a desk, allowing the magnet to hang from the desk.
4. Move the north pole of the second magnet next to the south pole of the hanging magnet. Record what happens.
5. Move the north pole of the second magnet next to the north pole of the hanging magnet. Record what happens.
6. Move the south pole of the second magnet next to the north pole of the hanging magnet. Record what happens.
7. Move the south pole of the second magnet next to the south pole of the hanging magnet. Record what happens.
8. Write a conclusion explaining why you think the magnets reacted the way they did.

N S N

See the electricity and magnetism skill sheets on pages 76 and 77.

Name ____________________

Kilowatt Calculations

Follow the directions below to find out how much electricity your family uses in a month.

Directions: Read the chart below. Estimate the number of hours each appliance is used in your home each month. Then, using the formulas shown, calculate the cost to use these appliances each month.

wattage x hours of use ÷ 1,000 = kilowatt-hours (kWh) per month

kWh per month x cost per kWh (about $0.14) = approximate cost per month

Appliance	Watts	Hours per Day	Hours per Month	kWh per Month	Approximate Cost per Month
television	120				
refrigerator	780				
incandescent lightbulb	100				
water heater	4,500				
dishwasher	1,200				
washing machine	400				
clothes dryer	2,500				
microwave oven	900				
regular electric oven	2,400				

Approximate total cost per month: ____________

Bonus: Reread the chart above. On the back of this sheet, list two appliances you feel you could use less each month. Then write two goals for conserving energy in your home. Share and implement your goals with your family.

Note to the teacher: Use with "Kilowatt Hours" on page 74.

Name____ Identifying electrical safety

Electrical Safety Checklist

Electricity is helpful to us in many ways. But it can also cause us problems if we are not careful with it. Use the checklist below to see how safe your home is.

A. Directions: Use the following checklist to check the electrical safety of your home. If the room meets each safety requirement, color the *P* for pass. If it does not pass, color the *F* for fail. In the last two columns, list additional rooms in your home that were not included.

	Kitchen	Family Room/ Living Room	My Bedroom	Parent Bedroom	Bathroom	________	________
1. All electrical cords in this room are in good condition and are not worn, cut, or broken.	P F	P F	P F	P F	P F	P F	P F
2. Electrical cords in this room are located out of high-traffic areas.	P F	P F	P F	P F	P F	P F	P F
3. Electrical applianc-es in this room are not located where they can easily come in contact with water.	P F	P F	P F	P F	P F	P F	P F
4. The electrical out-lets in this room do not have too many items plugged into them.	P F	P F	P F	P F	P F	P F	P F

B. Directions: Read each statement below and then decide if it is true or false based on your family's behavior.

5. During a thunderstorm, my family unplugs the television and computer unless the items are connected to a surge protector. T F

6. Members of my family never use metal utensils to remove food from a toaster.

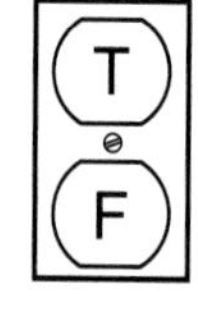

7. Members of my family never use electrical appliances in the bathtub.

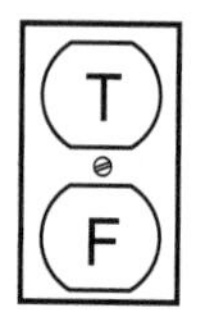

8. My family replaces worn electrical cords.

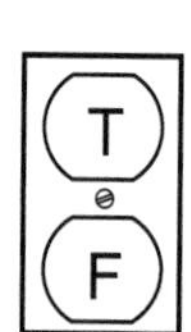

Bonus: For each *F* you filled in above, write a sentence explaining what changes you could implement to make your house safer.

Name ______________________________

Electrifying Experiments

Follow the directions in each box to complete the experiments.

Static Strips

Materials: two 1" x 10" lengths of newsprint

Procedure:

1. Hold the strips of newsprint together at one end with one hand.
2. Put the middle finger of your other hand between the strips and place the surrounding fingers on the outside of the strips.
3. Squeeze your fingers together and pull your hand down the strips.
4. **Observations:** What did you see happen?

   ____
   ____
   ____
   ____

5. **Conclusion:** Why do you think it happened?

   ____
   ____
   ____
   ____
   ____
   ____

Stop Pushing Me!

Materials: 2 like magnets, centimeter ruler

Procedure:

1. Hold the two magnets close together to find the sides that attract each other.
2. Hold one magnet at the 0 centimeter mark. Hold the other magnet at the opposite end of the ruler.
3. Move the second magnet closer to the first magnet until you feel them attracting, or pulling, each other. Record this measurement.
4. Again hold magnet 1 at the 0 centimeter mark. Turn magnet 2 around so the magnets repel each other.
5. Move the second magnet closer to the first magnet until you feel them repelling, or pushing, each other. Record this measurement.
6. **Observations:** Was the pull or push of the magnets greater, or were they about the same? ______________
7. **Conclusion:** Why do you think the magnets reacted the way they did?

   ____
   ____

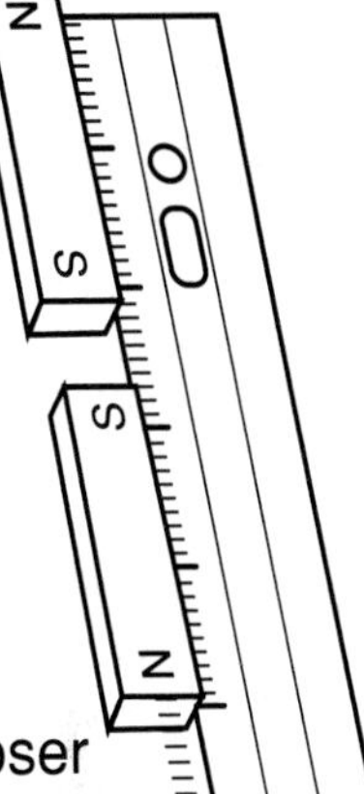

Magnet Muscle

Materials: 3 magnets of different sizes and shapes, 25 paper clips

Procedure:

1. Pile the paper clips on a desk.
2. Place the first magnet over the pile of paper clips. Record the number of paper clips the magnet picks up.
3. Repeat Steps 1–2 for each of the remaining magnets.
4. **Observations:** Which magnet picked up the most paper clips?

   ____

5. **Conclusion:** Why do you think that magnet was able to pick up more paper clips than the other magnets?

   ____
   ____

Magnet Description	Number of Paper Clips Lifted
1.	
2.	
3.	

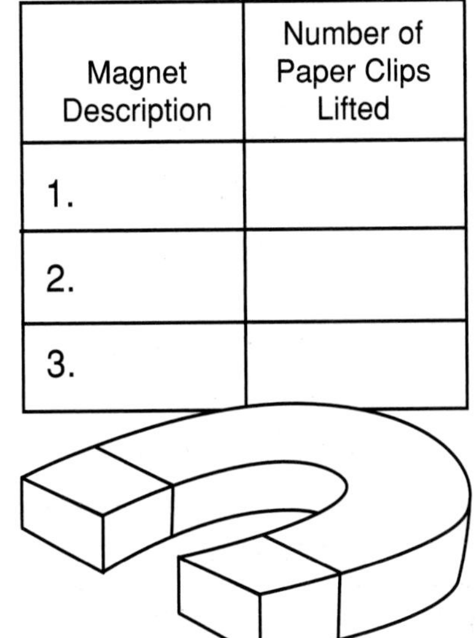

Note to the teacher: Plan on three 20-minute sessions to complete these experiments..

Rocks & Soil

Rock Vocabulary Study Guide

Have each student make one of these guides and keep it handy throughout your study of rocks and minerals. When an important vocabulary word is used, direct each child to write the word on the guide's rock pocket and then write the matching definition on the corresponding blank on the rock slide.

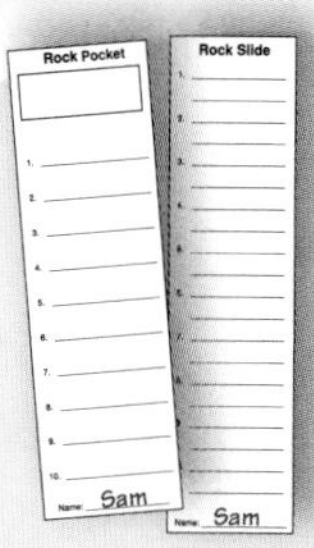

Figure 1

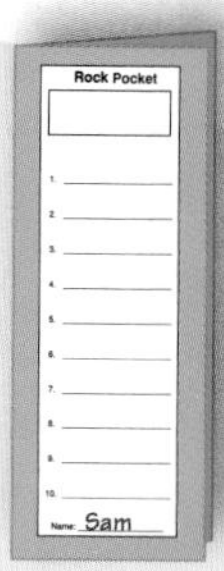

Figure 2

Materials for each student:
copy of page 80
6" x 9" sheet of construction paper
tape

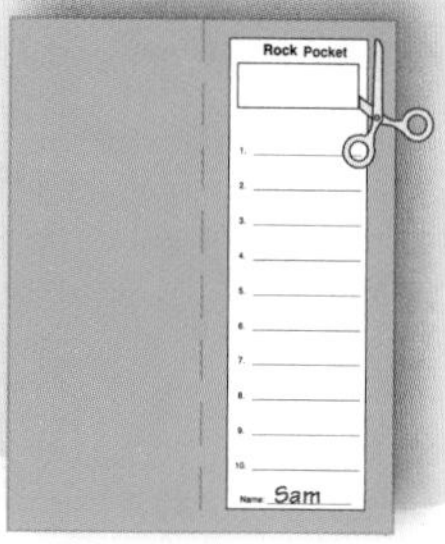

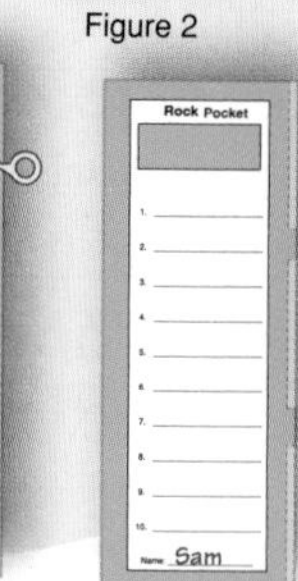

Figure 3

Figure 4

Steps for students:

1. Cut out the patterns from page 80. Write your name on each cutout (Figure 1).
2. Fold the construction paper in half lengthwise; then glue the rock pocket to one side (Figure 2).
3. Unfold the construction paper and cut out the window at the top of the rock pocket, cutting through the rock pocket and the construction paper (Figure 3).
4. Refold the construction paper, taping the right edges together (Figure 4). Do not tape the top or bottom.
5. Insert the slide into the construction paper pocket, sliding it until the space for definition 1 can be seen in the window (Figure 5).
6. Write vocabulary words and their corresponding definitions on your rock pocket and slide as your teacher directs (Figure 6).

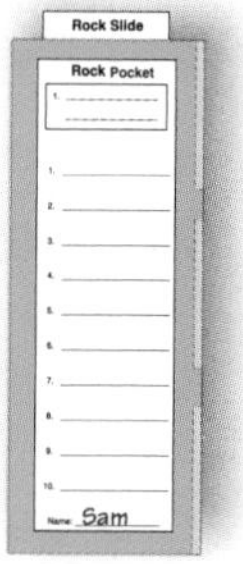

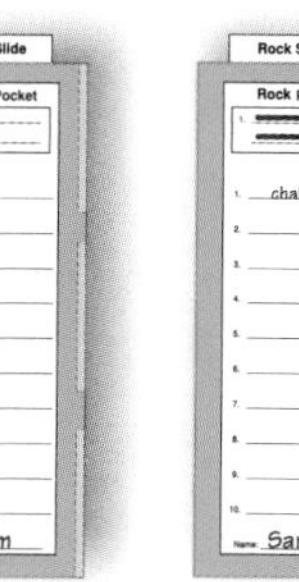

Figure 5

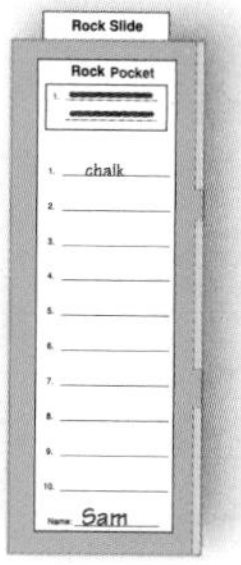

Figure 6

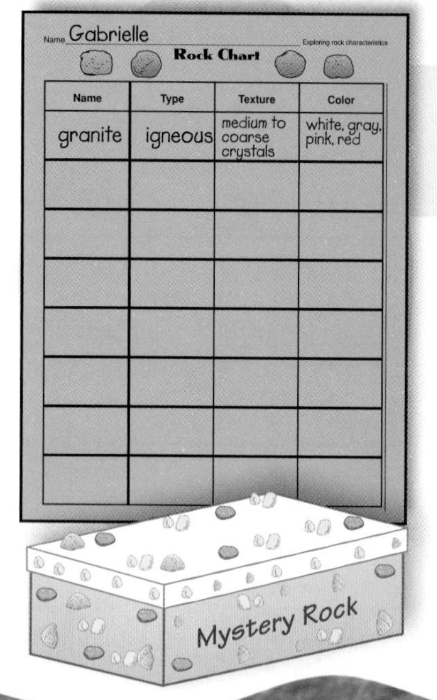

Name That Rock!

Materials:
class supply of copies of page 81
samples of 8 different varieties of rocks
decorated box labeled "Mystery Rock"

Day 1: Introduce your students to the qualities of different rock varieties. Pass around a rock sample. After each student has seen the rock, invite volunteers to tell about the texture, type, and color of the rock. Have students list the features in one row of their copies of page 81. Guide each student to determine the name of the featured rock and then write it in the appropriate column of her chart. Repeat this process with each rock sample.

Day 2: Review the information on the students' charts with them. Then secretly place one of the rocks in the box. Invite a student to ask a yes-or-no question about the rock, such as "Is the rock metamorphic?" or "Is it green?" Using her chart as a guide, have that child guess the name of the rock. If she is incorrect, continue with additional students. Once the mystery rock has been named, invite the student who correctly identified it to choose the next rock to go in the box and then answer her classmates' questions.

Rock-Type Road Maps

Materials for each group of students:
copy of page 82
access to reference materials

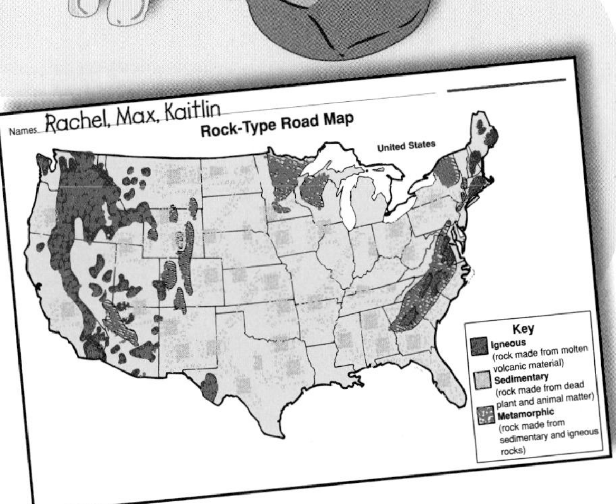

Review with students the three main types of rocks: igneous, sedimentary, and metamorphic. Discuss how these rocks can be found in different areas of the United States (see the map shown). Next, divide students into groups and provide each group with a copy of page 82. Direct each group to research where in the United States each rock type can be found. Have each group color the map key and then color its map accordingly.

Did You Know?
Most rocks are aggregates, or combinations of one or more minerals.

Metamorphic Marshmallows

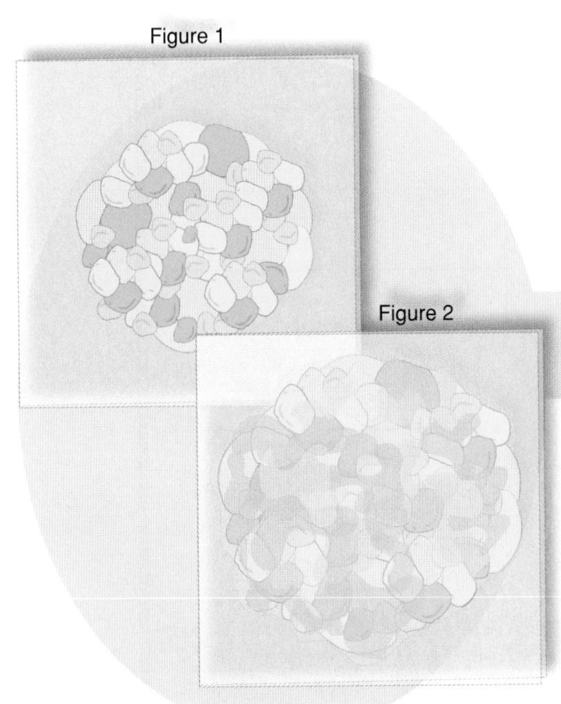

Explain to students that igneous and sedimentary rocks can be changed into metamorphic rocks by the extreme pressure and heat deep within the earth. This heat and pressure can cause minerals from these rocks to move, to be squeezed into new sizes and shapes, or to become completely new minerals. Direct students to follow the steps below to make a metamorphic rock of their own. Afterward, discuss what happened to the marshmallow pieces and guide students to realize how this is similar to the process igneous and sedimentary rocks go through in becoming metamorphic rocks.

Materials for each student:
20 small, colored marshmallows
two 6" squares of waxed paper

Steps for students:
1. Tear the marshmallows (representing igneous and sedimentary rocks) into small pieces.
2. Pile the marshmallow pieces in the center of one waxed paper square.
3. Lay the second waxed paper square on top of the pile.
4. Use your palm to press down hard on the marshmallow pieces for several seconds (Figure 1).
5. Place the waxed paper and marshmallows outdoors in a sunny spot for about 30 minutes.
6. Peel the waxed paper off and observe the changes in the marshmallows (Figure 2).

Rock Pocket and Rock Slide Patterns

Use with "Rock Vocabulary Study Guide" on page 78.

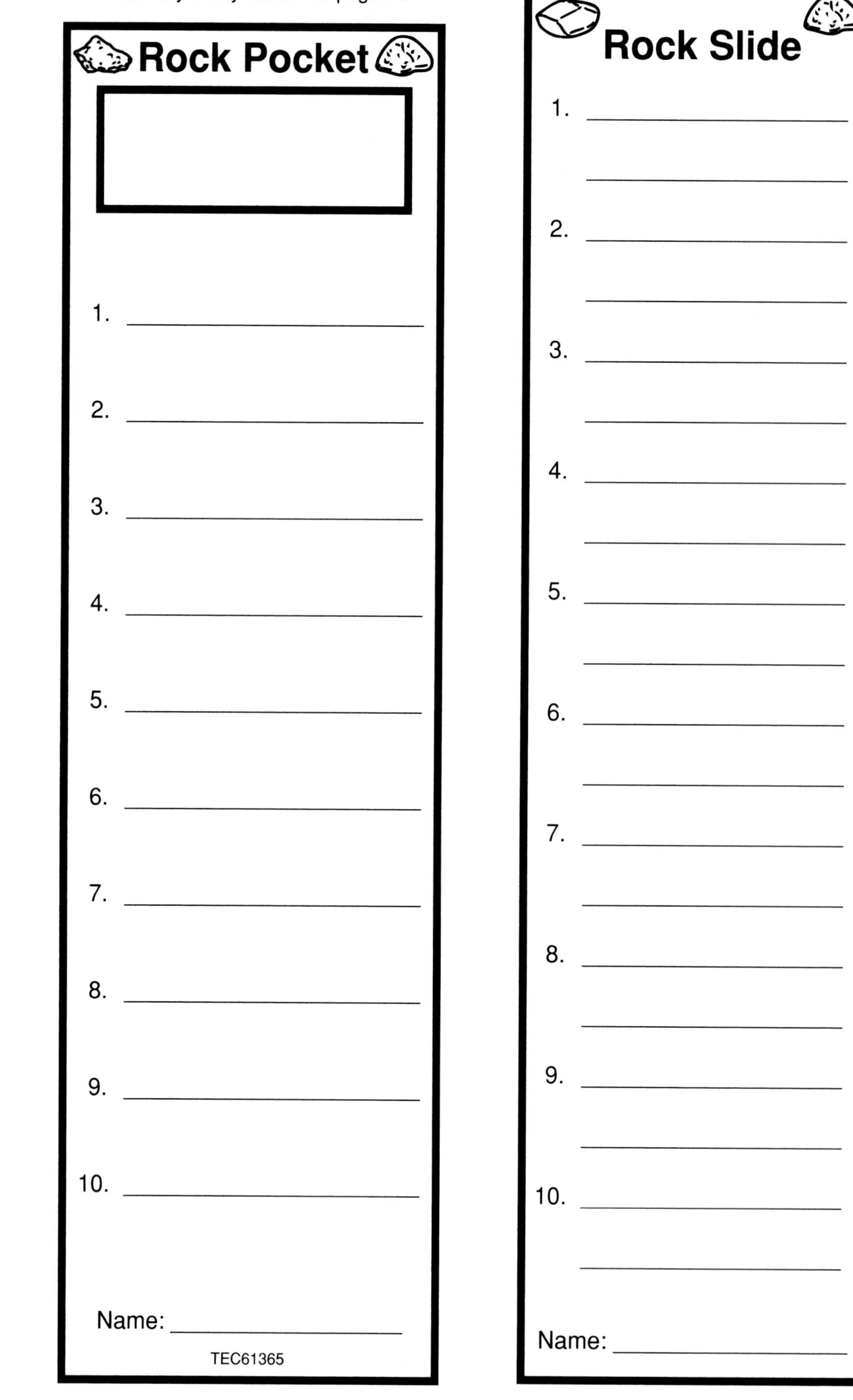

Name ____________________

Rock Chart

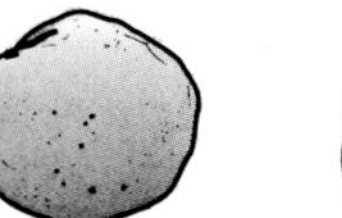

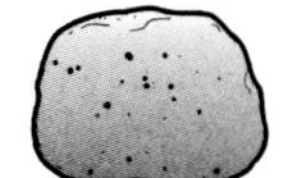

Name	Type	Texture	Color

Note to the teacher: Use with "Name That Rock!" on page 78.

Name ____________________

Rock-Type Road Map

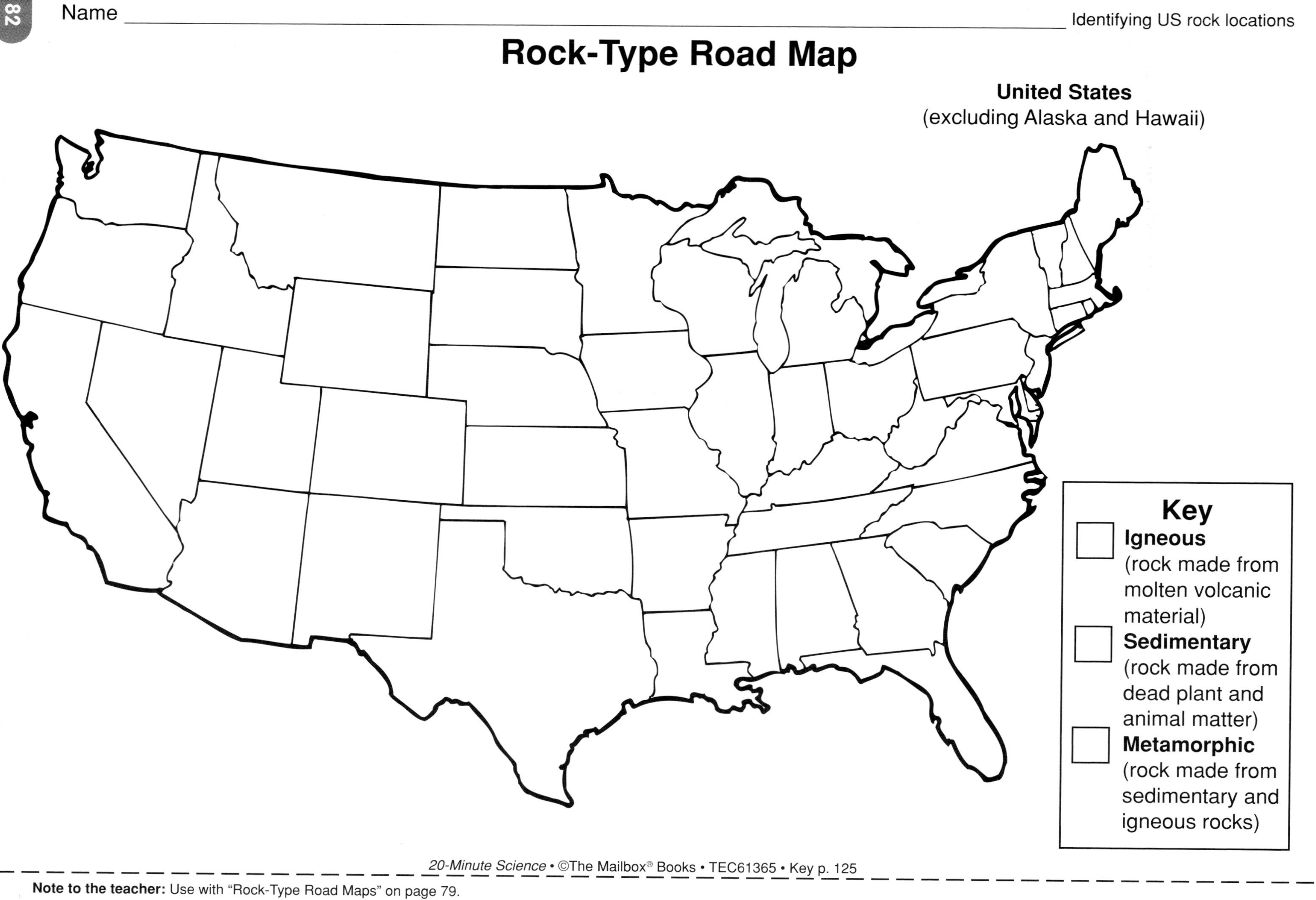

Note to the teacher: Use with "Rock-Type Road Maps" on page 79.

Weather & Climate

Weather Vane

Materials for each small group:
copy of page 92 for each student
brick
chopstick
small length of poster board
modeling clay
pen cap
four blank index cards

Note: You will also need one compass to use with the entire class.

Have students make these projects to monitor the direction from which the wind blows. Invite one student in each group to cut an arrow from the poster board and tape a pen cap to its middle. Ask a different group member to press a tiny lump of clay on the arrow point as shown. Then direct a student in each group to use clay to attach one end of the chopstick to the brick and then slide the pen cap on the top of the chopstick. Ask a member from each group to label each card "N," "S," "E," or "W." Help each group take its materials outdoors. Using the compass, guide each group's members in taping their directional cards to the appropriate sides of their brick.

Observation: Each day for two weeks, have each student check his group's weather vane to see the direction from which the wind is blowing. Then ask each child to record the wind direction on his copy of page 92.

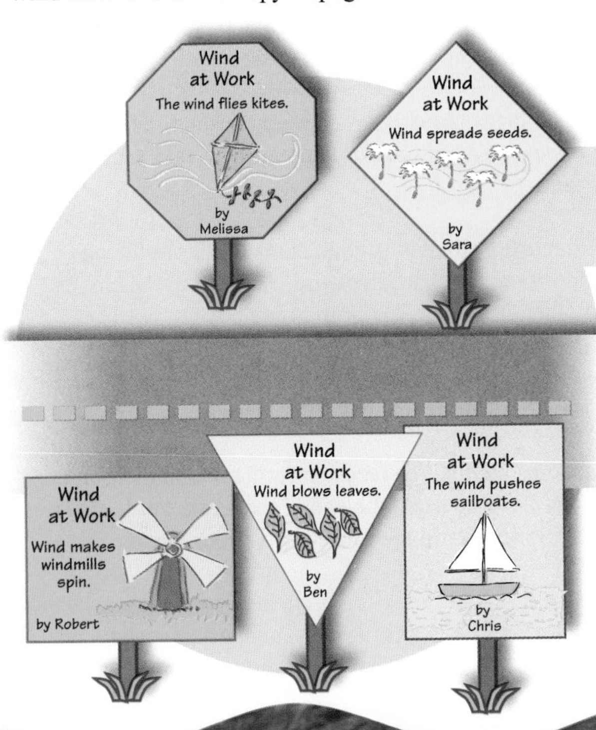

Wind at Work

Materials for each student:
sheet of construction paper
brown construction paper strip

Tell students that wind power helps people and the environment. Then help them brainstorm different ways the wind does work, such as spreading seeds, flying kites, pushing sailboats, providing power, and bringing rain. List their ideas on the board. Then have each student cut a road sign shape from her construction paper. Direct each child to glue the paper strip to the bottom of her sign so that it looks like a pole. Next, have each child write "Wind at Work" at the top of her sign. Below the title, ask each student to write on her sign an example of a job that wind does. Then have her add an illustration. Post the completed signs on a display similar to the one shown.

Wind Names

Materials:
access to reference materials about wind

Share with students different wind words, such as *doldrums* (a belt of calm, light breezes or sudden squalls near the equator, mainly over oceans), *trade wind* (a strong wind that blows toward the equator from the northeast or southeast), *horse latitudes*, (regions noted for their lack of wind), and *chinook* (a warm, dry wind that blows down the eastern slopes of the Rocky Mountains, usually in winter and early spring). Then pair students and challenge each duo to find a type of wind. Have each twosome use reference materials to research the origin of their chosen wind word and then write a paragraph about it.

Maggie and Tomas
Horse Latitudes
Horse latitudes are regions located at about 30° north and 30° south latitude. They are known for their lack of wind. Sailors named these regions horse latitudes first probably because many horses died on sailing ships delayed because of lack of wind. Most of the world's major deserts lie at horse latitudes.

Cloud Collection

This activity shows students the relationship between certain types of clouds and precipitation.

Cloud Collection Chart					
	Monday	Tuesday	Wednesday	Thursday	Friday
Cloud Illustration					
Cloud Type	Cumulus	Cumulonimbus	Cirrus	Cumulus	Altocumulus
Weather Illustration					
Weather Description	mostly sunny	hard rain	sunny	mostly sunny	light rain or drizzle

Materials for each student:
copy of page 93
cloud collection chart with headings like the one shown

Day 1: Review with students the cloud types and terms from the top of page 93 *(cumulus, stratus, cirrus, nimbus,* and *alto).* Then direct each child to complete his copy of page 93.

Day 2: Have each child make a cloud collection chart similar to the one shown. After each child makes his chart, invite him to observe the clouds and sketch the type of cloud he sees in the appropriate box on his chart. Then have him identify and label the cloud he drew, referring to his copy of page 93. Also have each student make a simple drawing—such as a sun, a snowflake, or raindrops—to illustrate that day's weather.

Observation: Have students continue to record information on their cloud collection charts each day for a week. At the end of the week, guide students to understand the types of precipitation produced by various clouds.

Follow-up: Have each child review the information he recorded on his cloud collection chart. Then have him write a paragraph summarizing what he learned about different types of clouds and precipitation.

Shifting Fronts

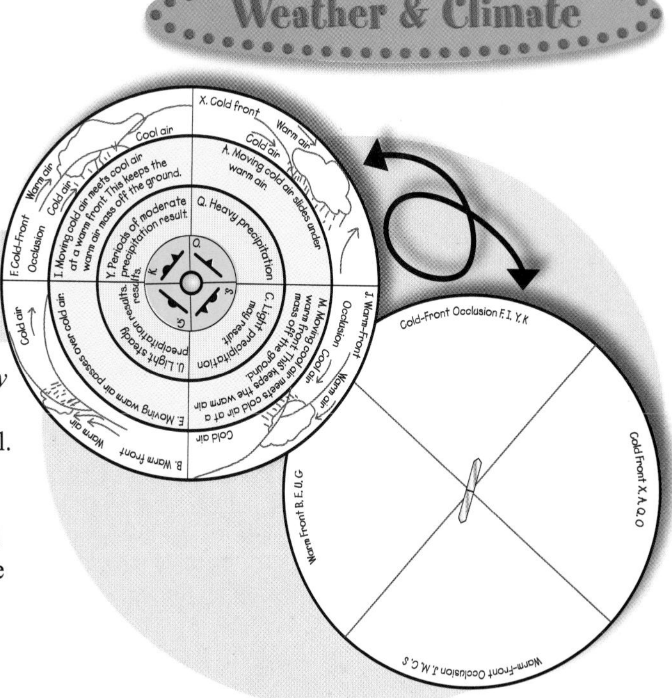

Materials for each student:
copy of pages 94 and 95
brad fastener

Day 1: Review with students the definition of a front *(a boundary between air masses of different temperatures)*. Then have each student follow the directions on page 94 to make a weather wheel.

Day 2: Demonstrate for students how to align the four circles to show the characteristics of each type of front. Discuss the frontal information for each type of front. Encourage each student to use his wheel as he continues his study of weather fronts.

Hailbob's
Bad Day
My name is Hailbob. This morning I was just a tiny ice crystal at the top of a puffy cumulonimbus cloud. The sun was shining and soon millions of other ice crystals and water droplets had crowded into my cloud. It began to get windy and everyone was pushing and shoving. I was thrown up and down and water droplets began to freeze to my skin. I was getting fatter and heavier by the minute. Suddenly I began to fall. I screamed for help but my friends were falling too! Down, down, down I dropped. Smash! Crackle! Ouch! I hit the window of a red truck.

Never a Dry Moment

A Chilly Childhood

Developing Precipitation

Materials for each student:
9" x 12" sheet of construction paper

Review with students how rain, snow, and hail form. Next, have each student choose a type of precipitation. From construction paper, have him make a cutout of the precipitation's shape. Direct each child to pretend to be his chosen type of precipitation developing in a cloud. Then have him write a story on his cutout to describe how he was formed.

How Does Precipitation Form?

- *Rain* forms when tiny droplets combine, making larger drops until they are too heavy to stay in the cloud.
- *Snow* forms when water droplets freeze into icy crystals in the cloud. As these crystals fall, they collide and stick together to form snowflakes.
- *Hail* forms when ice crystals are tossed up and down in a cumulonimbus cloud and are coated with layers of freezing water.

Blowing in the Wind

Materials for each pair of students:
copies of page 96 and page 97

Explain to students that wind speed is a major factor in the strength of a storm and that the Beaufort Scale is a tool used to categorize these speeds. Pair students and have each child look at a copy of page 96 as you discuss the Beaufort Scale. Next, have each pair cut out the game cards from page 97 and lay them out face down. Direct each duo to play a game similar to Concentration. In this game, a match is made when an observation card and a matching Beaufort Scale card are flipped. (Have students use page 96 to help them determine matches.) The child with more cards at the end of the game (or playing time) is the winner.

Beaufort Wind Scale

Number	Speed/MPH	Description	Observation
0	less than 1	calm	smoke rises vertically
1	1–3	light air	smoke drifts slowly
2	4–7	light breeze	leaves rustle; wind felt on face
3	8–12	gentle breeze	leaves and small twigs move
4	13–18	moderate breeze	small branches move
5	19–24	fresh breeze	small trees sway
6	25–31	strong breeze	large branches sway
7	32–38	moderate gale	whole trees sway; difficult to walk against wind
8	39–46	fresh gale	twigs break off trees
9	47–54	strong gale	shingles blown off roof
10	55–63	whole gale	trees uprooted
11	64–73	storm	widespread damage
12–17	74 and above	hurricane	extreme damage

That tree just pulled right out of the ground!

Beaufort Scale 10

Measuring Air Pressure

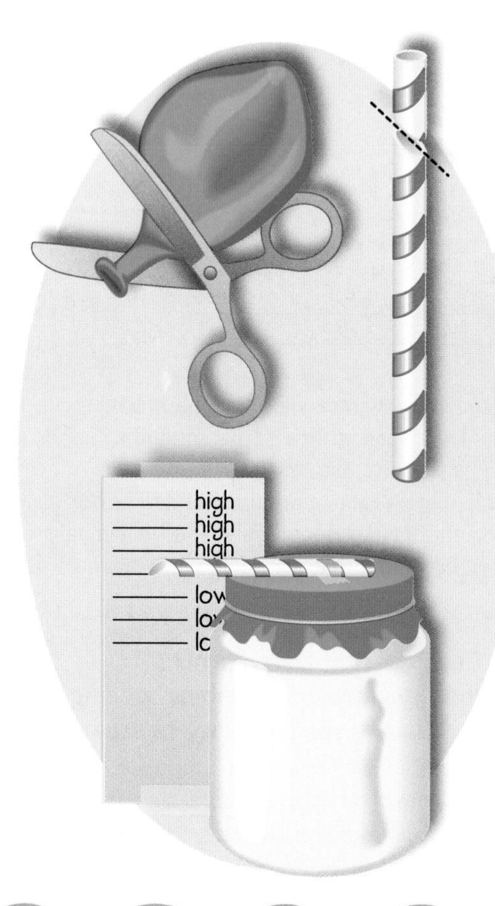

Tell students that understanding changes in air pressure helps predict when a storm will occur. Explain that when air pressure is high, the weather is often clear; when air pressure is low, the weather is often stormy. Then group students and lead them through the following steps to make barometers.

Materials for each group of students:
plastic or glass jar
piece of cardboard
drinking straw
balloon
strong rubber band
tape
ruler

Steps for students:

1. Cut off the balloon nozzle. Gently stretch the larger opening across the mouth of the jar.
2. Secure the balloon in place with a rubber band.
3. Cut the straw at an angle to make a pointed end. Tape the uncut section of the straw to the flat balloon surface.
4. Hold the cardboard next to the jar. Where the straw points, mark the cardboard. Then, at the mark, draw a horizontal line across the cardboard.
5. Draw three lines above and three lines below the drawn line, each one-fourth inch apart. Label the top three lines "high" and the bottom three lines "low."
6. Place the barometer on a shelf or table near a wall. Tape the cardboard piece to the wall so that the pointed end of the straw aligns with the middle line.

Observation: Each day for a week (or longer if there hasn't been any rain in your area), have each group record the date, whether the air pressure is low or high, and the outside weather conditions. Lead students in a review of the relationship between air pressure and the type of weather it signals.

Did You Know?
Barometers are instruments used to measure atmospheric pressure.

Flash! Boom! Bang!

This demonstration shows students the relationship between thunder and lightning. You will need two volunteers and a small paper bag.

Steps:

1. Have one volunteer (lightning) stand near the classroom light switches. Give the other volunteer (thunder) the paper bag and have him stand at the front of the classroom. Turn off the lights.
2. Direct the lightning to flip the light switches on and off once for a bolt of lightning.
3. Join students in slowly counting seconds: one thousand one, one thousand two, and so on. At the same time, the thunder blows inside the bag and traps a giant breath of air.
4. When the count reaches ten seconds, the thunder hits the bottom of the air-filled bag, making a thunder-like sound.
5. Tell students that even though thunder and lightning occur simultaneously, the sound of thunder takes much longer to reach us than the sight of light.

Did You Know?
To determine how far away lightning is, count the number of seconds between the sight of lightning and the sound of thunder and then divide by five (because sound travels one-fifth of a mile per second.)

Hurricane Spin

Explain to students that a hurricane has four parts and, in each part, the wind speeds are very different: the *eye*, calm; the *eye wall* or *inner hurricane*, about 120–150 mph; the *middle hurricane*, about 74 mph; and the *outer hurricane*, about 40 mph. Then take students to an open, flat area, such as a gymnasium or a ball field. Divide students into groups of four and have each group form a line, standing an arm's length apart. Instruct each child to place her hands on the shoulders of the students next to her. (See illustration.) Designate the student at one end of each line to represent the eye of a hurricane and have him stand calmly apart from the line. Direct the remainder of each line to begin walking around in a circle around the eye. Stop students and explain that, in a hurricane, the winds spin around the eye but the eye stays very still in comparison. If desired, repeat the activity allowing different children to take turns being the eye.

Weather-Forecasting Tools

Materials for each group:
access to reference materials
piece of tagboard

Day 1: Challenge students to research instruments used in weather forecasting (see the list). Divide students into groups and assign each group a different weather instrument to research. Ask each group to illustrate its weather tool and write a brief description of how it is used to forecast weather.

Day 2: Have the groups complete their research and share their projects with the class as time permits.

Weather Instruments

thermometer	rain gauge
barometer	radar
weather vane	weather balloon
anemometer	weather satellite
hygrometer	ocean buoy

Long- and Short-Range Predictions

Materials:
recording chart like the one shown
Internet access to weather forecasts

Long-Range Forecast

MON	TUES	WED	THUR	FRI
45	50	39	34	32
26	33	35	22	20
Right On	Partially Correct			

Short-Range Forecast

MON	TUES
45	50
26	33
Right On	Right On

Track the accuracy of short-range and long-range weather forecasts for one week. To begin, record the long-range weather forecast for five days. Then record the short-range forecast for Day 1. Explain to students that a short-range forecast is typically for a period of 18 to 36 hours, while a long-range forecast is usually for a period of five to seven days. Point out that the long-range and short-range forecasts for Day 1 are the same because they were predicted on the same day.

Observation: On each of the following days, have students evaluate the accuracy of the previous day's forecasts and rate each one as either "Right on," "Partially correct," or "No way." Have them record each rating and the short-term forecast for the current day. At the end of the five days, compare with students the accuracy of the short-range and long-range forecasts and offer explanations for students' observations.

Keeping Up With Climate

Materials for each group:
copy of page 98
access to reference materials

Weather: the condition of the outside air at any given time and place
Climate: the average weather in one place over a period of time

Teach students about the climate characteristics of different cities in the United States. Set aside two or more 20-minute sessions for research.

Days 1 and 2:
Research—Verify that students understand the differences between climate and weather. Then pair students. Ask each pair to select a city in the United States and investigate its climate by completing a copy of page 98.

Follow-up—Invite each pair to share with the class what it learned about its chosen city's climate, including traveling and packing tips.

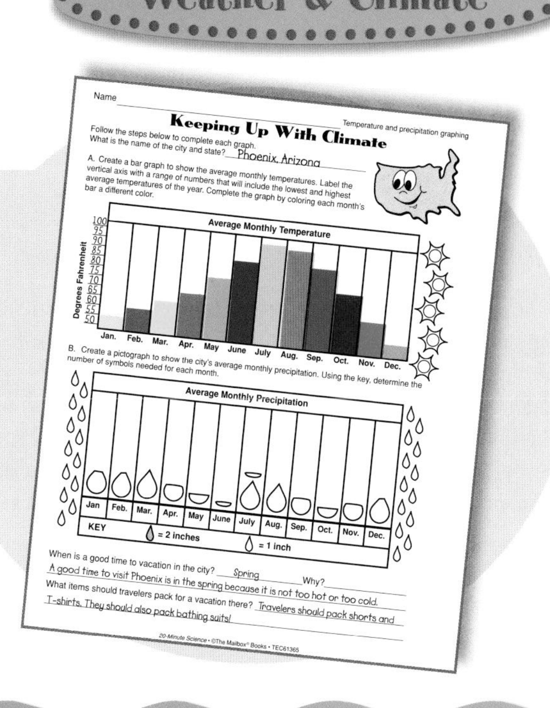

Name ____________

Temperature and precipitation graphing

Keeping Up With Climate

Follow the steps below to complete each graph.
What is the name of the city and state? Phoenix, Arizona

A. Create a bar graph to show the average monthly temperatures. Label the vertical axis with a range of numbers that will include the lowest and highest average temperatures of the year. Complete the graph by coloring each month's bar a different color.

B. Create a pictograph to show the city's average monthly precipitation. Using the key, determine the number of symbols needed for each month.

When is a good time to vacation in the city? Spring Why? A good time to visit Phoenix is in the spring because it is not too hot or too cold.

What items should travelers pack for a vacation there? Travelers should pack shorts and T-shirts. They should also pack bathing suits!

20-Minute Science • ©The Mailbox® Books • TEC61365

Did You Know?
Climatologists—scientists who study climate—mainly rely on temperature and precipitation averages to determine climate.

It's All in Your Latitude

Use this demonstration to introduce your students to one of nature's most important climate controls—latitude.

Materials:
desk lamp
2 outdoor thermometers
9" x 13" baking dish filled with sand or dirt

Steps:
1. Place the thermometers in the sand or dirt at opposite ends of the pan. Make sure both thermometers are facing the same direction.
2. Place the lamp near the first thermometer so the light is shining directly on it.
3. Observe the thermometers for 15 minutes. At five-minute intervals choose a different student to read the temperatures and record them on the board.
4. Discuss with students why the thermometer nearer the lamp became warmer. Explain that this demonstration illustrates why land nearer the equator often has higher temperatures. *(These areas receive more direct sunlight than areas farther from the equator.)*

Location, Location, Location

Materials for each student:
copy of page 99

Remind students of the relationship between climate and proximity to the equator. Also explain that scientists divide the earth into six major regions, called *biomes*, based on climate and plant and animal life. Next, ask students to listen carefully as you read aloud a biome clue from below. Challenge students to use the information revealed in the clue as they search their maps to find the location of the biome described. Ask a volunteer to share the location of the biome and identify the matching key symbol. Then have each student label the appropriate key symbol, color it a desired color, and color each section of the identified biome on his map. Repeat this process for each biome.

Biome Clues

I can be found in the coldest parts of the world. I am usually covered with snow. There are long winters and short, cool summers here. Because I am found in the most northern areas, it is too cool for trees to grow here. I am the tundra.

I am considered a subarctic forest. Trees can grow here even though I am located in the north and it is very cold. I begin where the tundra ends. I am the taiga.

I have a seasonal climate where there are warm summers and cold winters. My climate allows deciduous trees (trees that shed their leaves seasonally) and evergreen trees (trees that keep their leaves all year) to grow. I can be found across most of the eastern United States. I am the temperate forest.

I have a warm, wet climate. My climate supports a large variety of plant and animal life. I can be found in a large section of northern South America. I am the tropical rain forest.

I am usually located between deserts and humid, forest-covered areas. There are two types of me: steppes, which have short grasses, and prairies, which have tall grasses. I am found on six of the seven continents. I am the grassland.

I am a hot, empty, dry region. Even though there is little rainfall where I am located, plants like cacti thrive in my climate. My largest area is in the Sahara in northern Africa. I am the desert.

Earth & Space Science

Water–A Climate Moderator

Help students understand that locations found at similar latitudes *can* have different climates. On a map or globe, show students Boston, Massachusetts, (a coastal city) and Omaha, Nebraska, (an inland city). Explain that these two cities have similar latitudes and yet Omaha has cooler January temperatures and warmer July temperatures. Then use the following demonstration to show how water can act as a climate moderator because it heats and cools more slowly than land.

Materials:
2 metal loaf pans
2 outdoor thermometers
plastic wrap
soil
tape

Steps:
1. Fill one loaf pan three-fourths full of soil. Fill the other loaf pan three-fourths full of water.
2. Tape a thermometer on the inside of each loaf pan as shown.
3. Cover each pan with plastic wrap.
4. Place the pans outdoors in the shade. Note the temperature of each pan after one minute.
5. Move the pans to a sunny spot. After five minutes, note the temperature of each pan.
6. Return the pans back to the shade. Observe how the temperatures decline at different rates.

Elevation Station

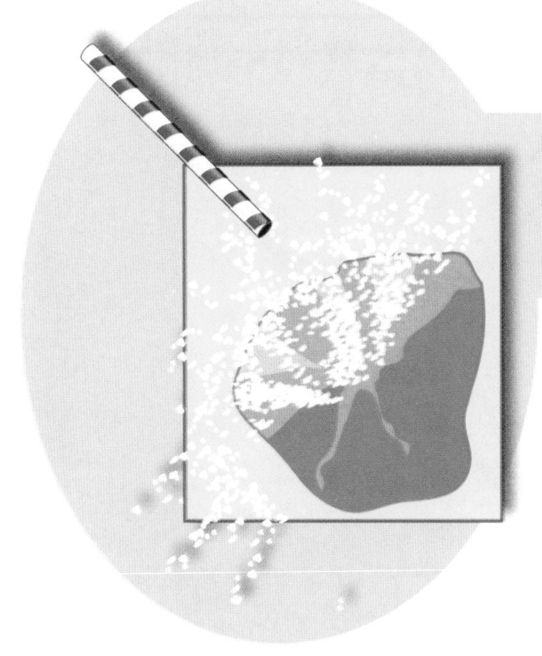

The role that mountains play in controlling climate becomes clear during this hands-on small-group activity.

Materials for each group:
12" x 18" sheet of construction paper
½ c. powdered sugar in a tall plastic cup
clay
straw for each group member

Steps for students:
1. Use the clay to form a mountain two to three inches high on your paper. Mark the mountain's windward side (facing the wind).
2. When it is your turn, cover one end of your straw with your finger and put the opposite end in the powdered sugar. Then, while at eye level to the paper, blow the sugar gently onto the windward side of the mountain.

Follow-up: Have students observe the sugar patterns on their mountains. Explain that the sugar represents the moisture that is released from the air as it cools and rises over a mountain. This shows that the windward side of a mountain receives more rainfall than the opposite (or *leeward*) side.

See the weather skill sheets on pages 100–102.

Name ______________________________

Wind Tracker

Each day for two weeks, use your weather vane to determine the direction of the wind. Then color one section of the bar on the kite for that direction. If on another day the wind blows from the same direction, color another section of the bar in that direction. See the sample below.

Sample
Shows the wind blowing two days from the east and four days from the southeast.

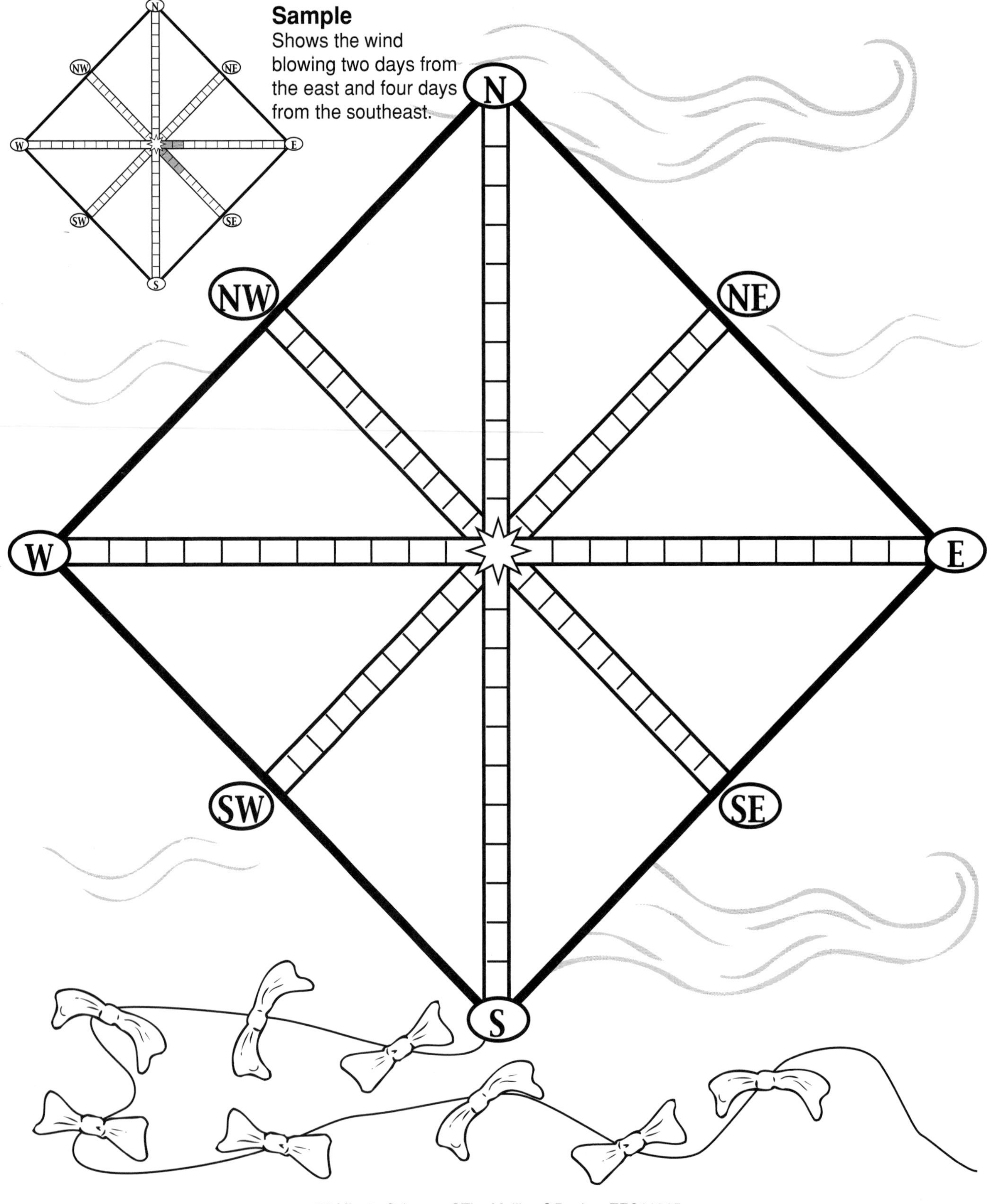

Note to the teacher: Use with "Weather Vane" on page 83.

Name ______________________________

Cloudy Combinations

Over 150 years ago, an English scientist named Luke Howard identified three basic types of clouds: **cumulus, stratus,** and **cirrus**. Howard added the terms **nimbus** to identify rain clouds and **alto** to describe higher clouds. Using a combination of these cloud types and terms, Howard identified several other types of clouds. Follow the directions below to learn more.

Directions: Read the information about the three basic clouds in the diagrams below. Then combine the Latin meanings to help you define each of the other types of clouds. Finally, sketch a picture of each cloud in the box.

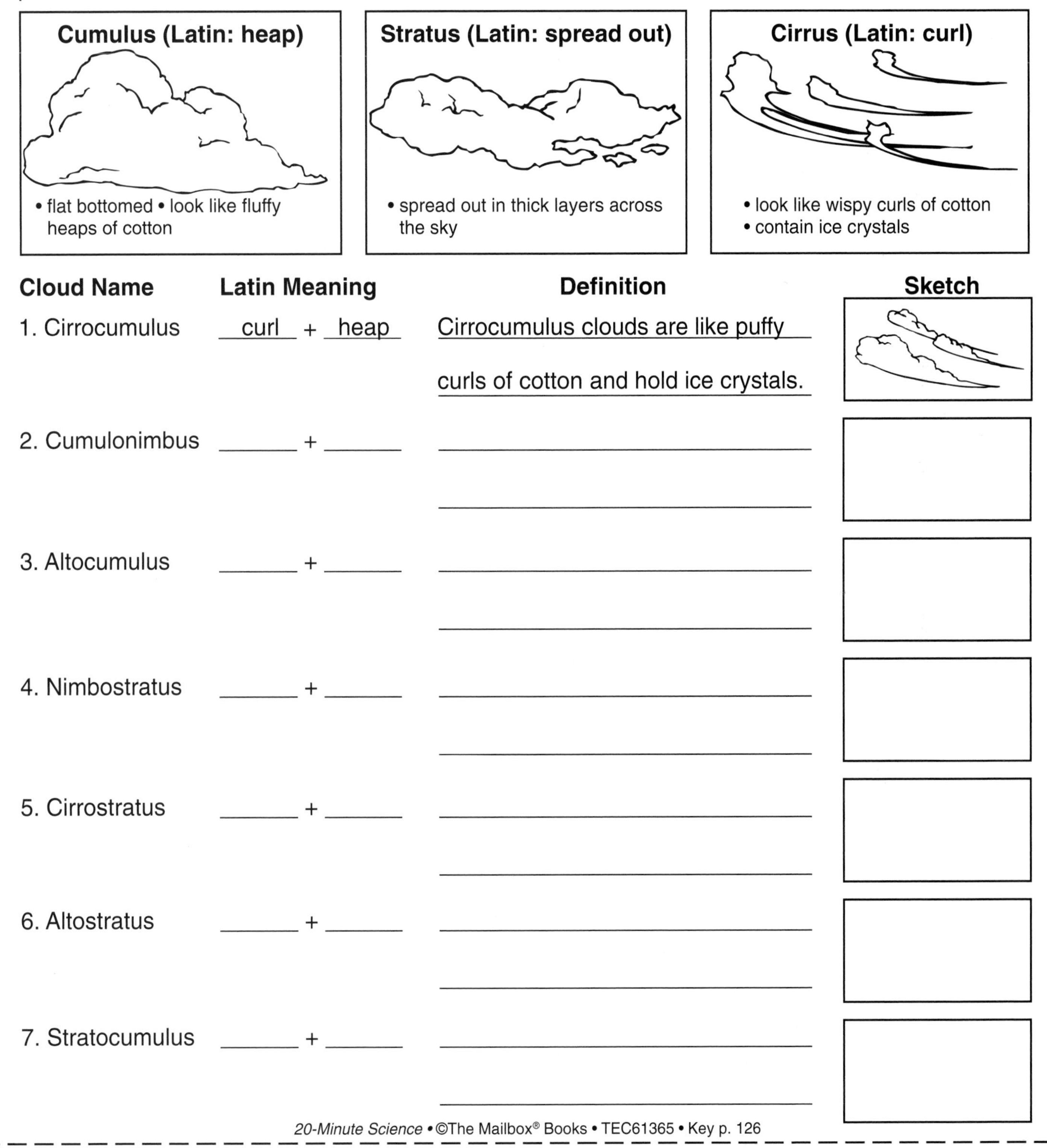

Cloud Name	Latin Meaning	Definition	Sketch
1. Cirrocumulus	curl + heap	Cirrocumulus clouds are like puffy curls of cotton and hold ice crystals.	
2. Cumulonimbus	______ + ______	______	
3. Altocumulus	______ + ______	______	
4. Nimbostratus	______ + ______	______	
5. Cirrostratus	______ + ______	______	
6. Altostratus	______ + ______	______	
7. Stratocumulus	______ + ______	______	

Note to the teacher: Use with "Cloud Collection" on page 84.

Weather Wheels

Use the following information and the steps below to make a weather wheel.

Fronts

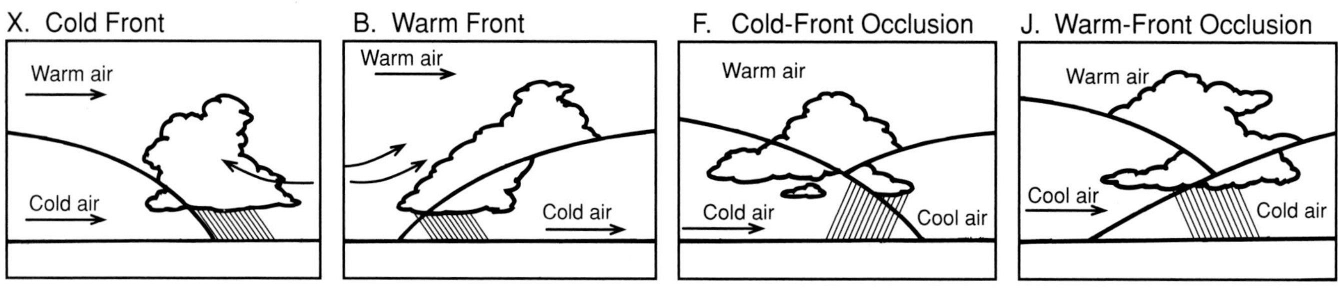

Description of Front

A. Moving cold air slides under warm air.
E. Moving warm air passes over cold air.
I. Moving cold air meets cool air at a warm front. This keeps the warm air mass off the ground.
M. Moving cool air meets cold air at a warm front. This keeps the warm air mass off the ground.

Weather Result

Q. Heavy precipitation results.
U. Light, steady precipitation results.
Y. Periods of moderate precipitation result.
C. Light precipitation may result.

Symbols

O.
G.
K.
S.

Steps:

1. Cut out the circle patterns on page 95.
2. At the top of each section on circle 1, write the term for each front and its matching letter. Then sketch the illustration for each front as shown above.
3. At the top of each section on circle 2, write the front description and its matching letter.
4. At the top of each section on circle 3, write the weather result and its matching letter.
5. For each section on circle 4, write the meteorological symbol and its matching letter.
6. Layer the circles atop each other as shown and secure them with a brass fastener.
7. On the back of the wheel, copy each answer code behind its matching front.

Note to the teacher: Use with "Shifting Fronts" on page 85.

Circle Patterns

Use with "Shifting Fronts" on page 85.

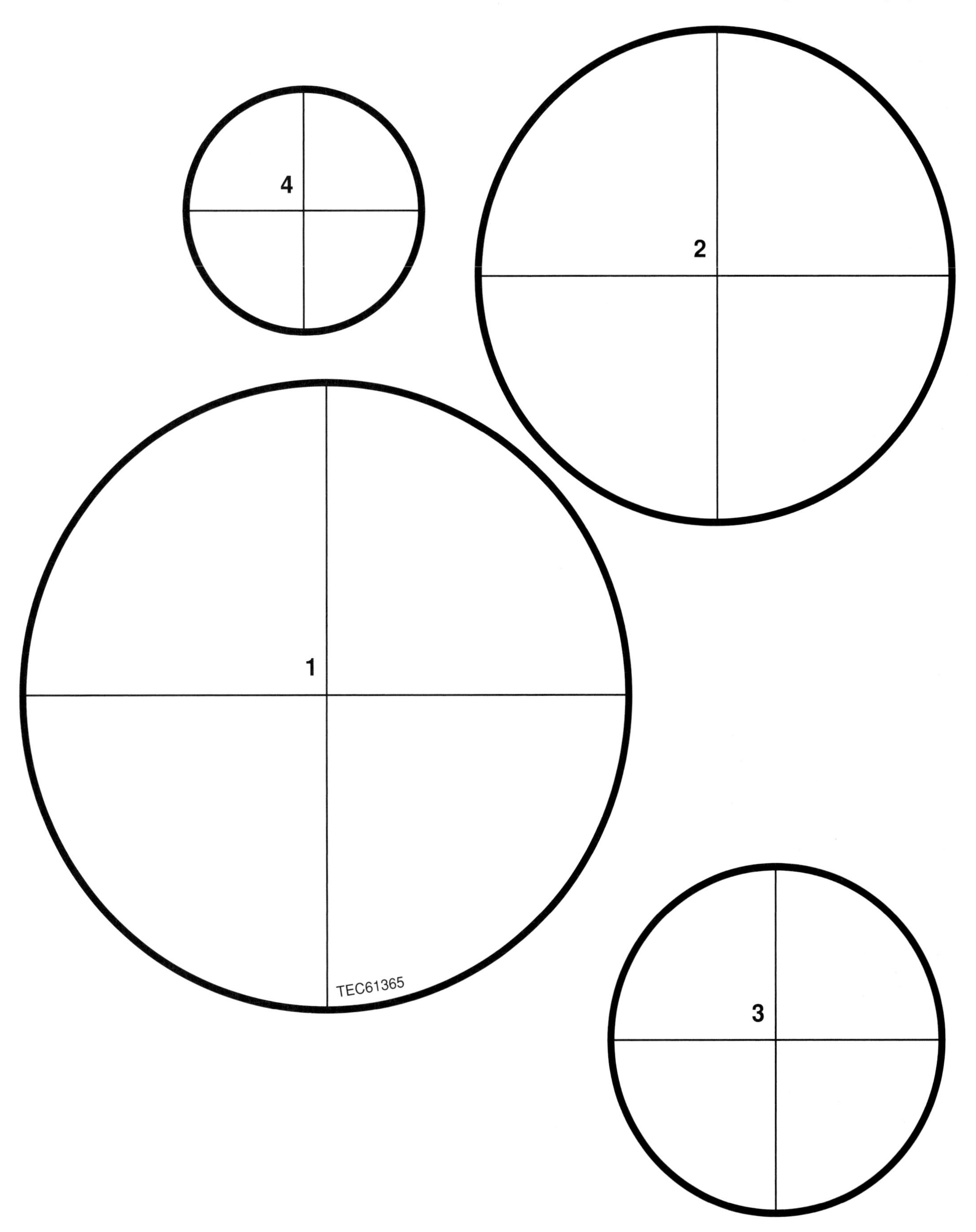

Beaufort Wind Scale Card

Use with "Blowing in the Wind" on page 86.

Beaufort Wind Scale

Number	Speed/MPH	Description	Observation
0	less than 1	calm	smoke rises vertically
1	1–3	light air	smoke drifts slowly
2	4–7	light breeze	leaves rustle; wind felt on face
3	8–12	gentle breeze	leaves and small twigs move
4	13–18	moderate breeze	small branches move
5	19–24	fresh breeze	small trees sway
6	25–31	strong breeze	large branches sway
7	32–38	moderate gale	whole trees sway; difficult to walk against wind
8	39–46	fresh gale	twigs break off trees
9	47–54	strong gale	shingles blown off roof
10	55–63	whole gale	trees uprooted
11	64–73	storm	widespread damage
12–17	74 and above	hurricane	extreme damage

Game Cards

Use with “Blowing in the Wind” on page 86.

Name ____________________

Keeping Up With Climate

Follow the steps below to complete each graph.
What is the name of the city and state? ____________________

A. Create a bar graph to show the average monthly temperatures. Label the vertical axis with a range of numbers that will include the lowest and highest average temperatures of the year. Complete the graph by coloring each month's bar a different color.

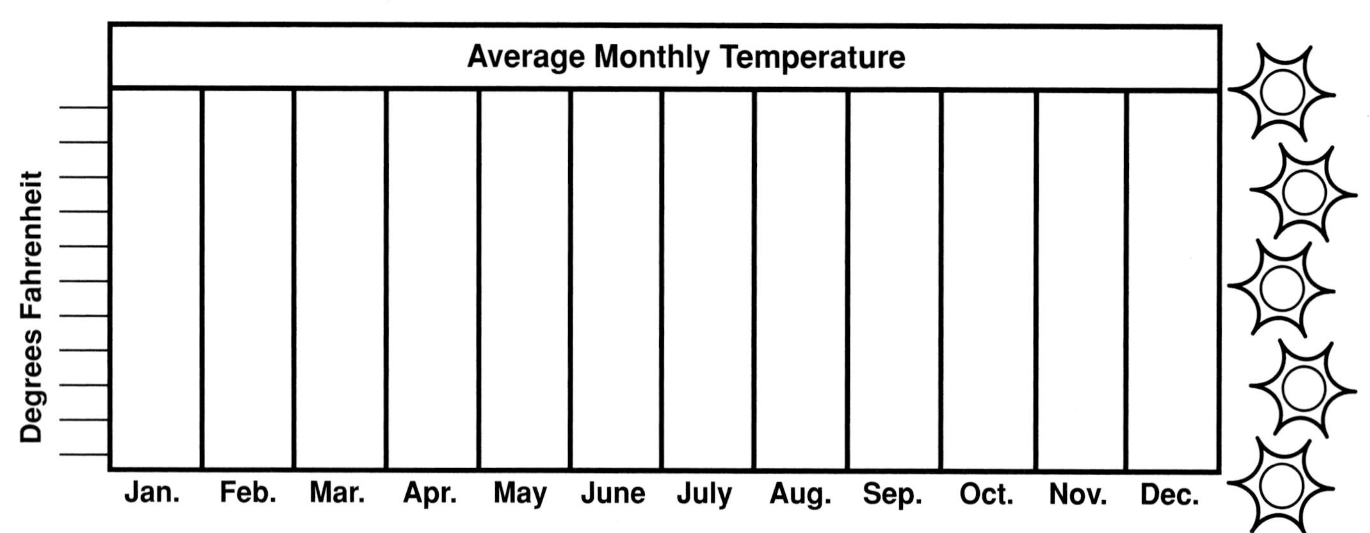

B. Create a pictograph to show the city's average monthly precipitation. Using the key, determine the number of symbols needed for each month.

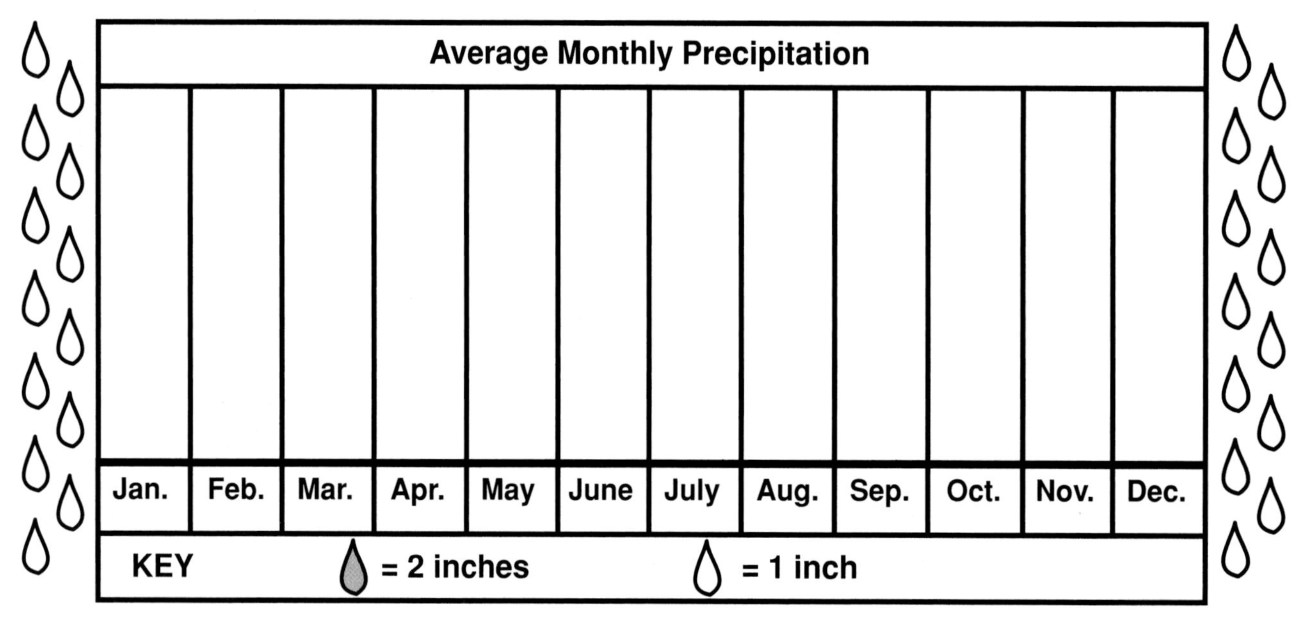

When is a good time to vacation in the city? ____________________ Why? ____________________

What items should travelers pack for a vacation there? ____________________

 Note to the teacher: Use with "Keeping Up With Climate" on page 89. Provide each student with an almanac and crayons or markers.

Name ____________________

Learn the Locations

The earth is divided into six major land regions called *biomes.* Each region's climate is different. The differences in climate affect the kind of plant and animal life found in each region. Listen as your teacher reads each biome clue; then label the key and color the correct area(s) on the map.

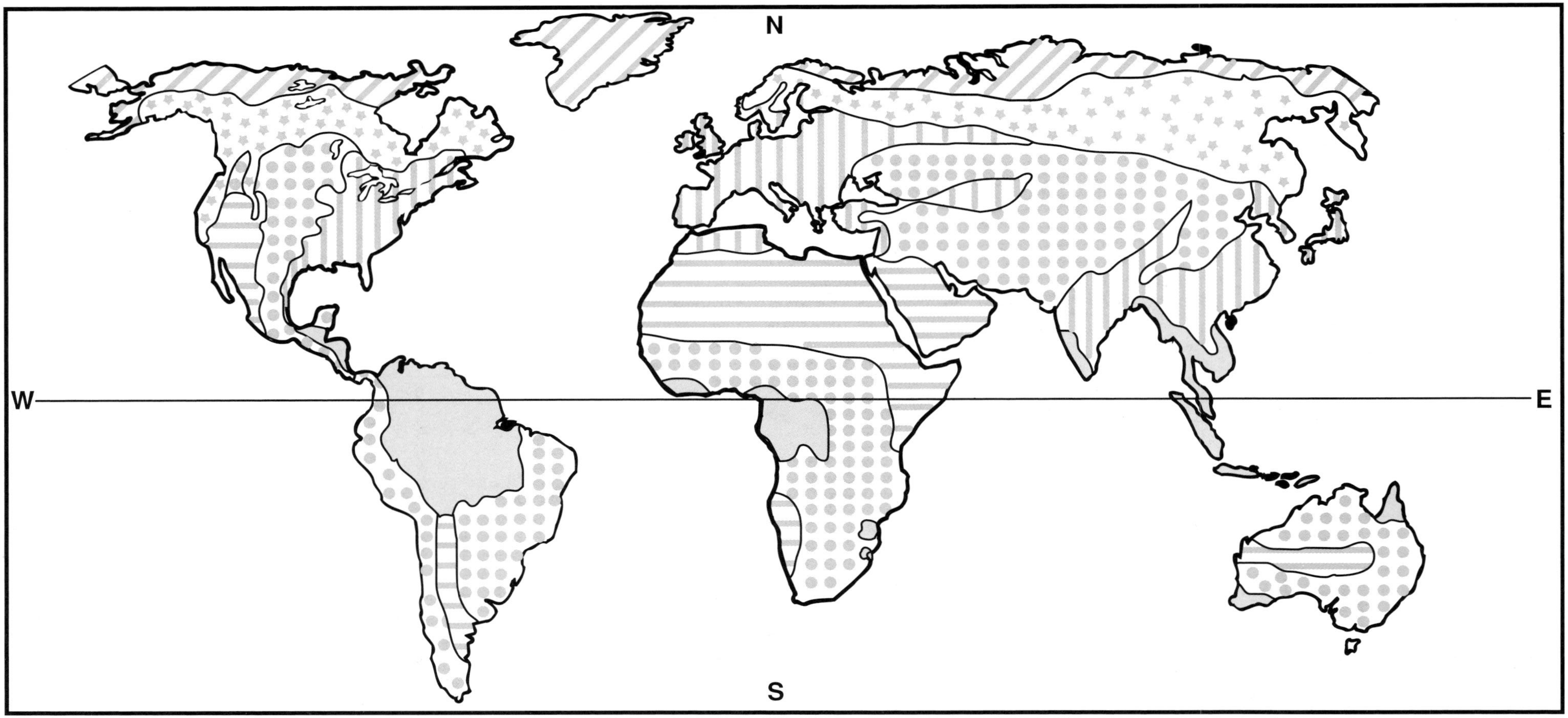

N

W — E

S

Biomes

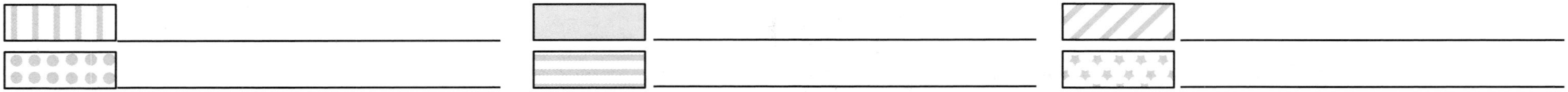

Note to the teacher: Use with "Location, Location, Location" on page 90.

Name ________________________________

Twistin' in the USA

Tornadoes can occur just about anywhere in the world. Most tornadoes in the United States occur in an area called Tornado Alley.

1. Choose one color for each range of numbers in the key using a different-colored pencil or crayon.
2. Use the key and the table below to color the map according to the average number of tornadoes occurring in each state each year.
3. Answer the map questions at the bottom of the page.

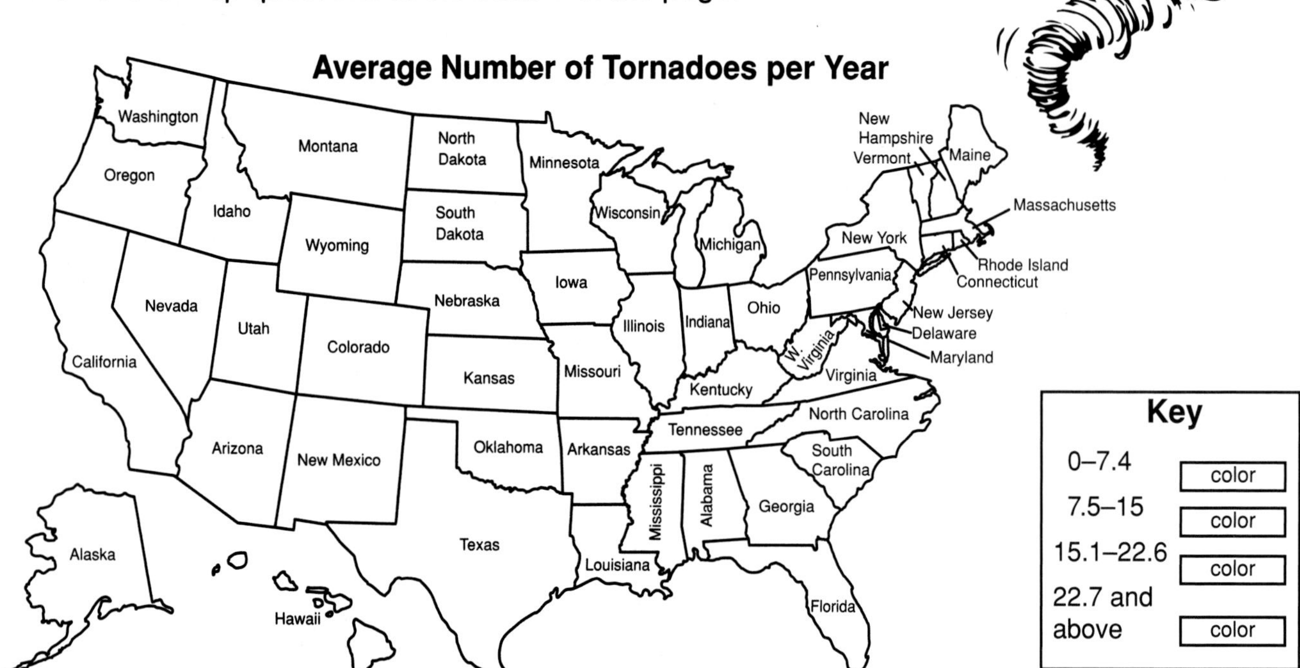

Average Number of Tornadoes Annually

State	Number	State	Number
Alabama	23	Montana	5
Alaska	0	Nebraska	36
Arizona	4	Nevada	1
Arkansas	21	New Hampshire	2
California	4	New Jersey	3
Colorado	24	New Mexico	9
Connecticut	1	New York	5
Delaware	1	North Carolina	14
Florida	52	North Dakota	20
Georgia	21	Ohio	16
Hawaii	1	Oklahoma	47
Idaho	2	Oregon	1
Illinois	27	Pennsylvania	10
Indiana	23	Rhode Island	0.23
Iowa	35	South Carolina	10
Kansas	36	South Dakota	28
Kentucky	10	Tennessee	12
Louisiana	27	Texas	137
Maine	2	Utah	2
Maryland	3	Vermont	1
Massachusetts	3	Virginia	6
Michigan	18	Washington	1
Minnesota	19	West Virginia	10
Mississippi	26	Wisconsin	21
Missouri	27	Wyoming	11

Questions:

1. What states do you think could be included in Tornado Alley? Why? ______________________________
____
____

2. Which state has a high number of tornadoes each year but is not part of Tornado Alley?
____

3. Think about the geography of the states in Tornado Alley. Are they mountainous, plains, or coastal? Do you think the type of geography has anything to do with the number of tornadoes a state gets? Why or why not?
____
____
____

Name ______________________________

Tracking Hurricane Windy

Track Hurricane Windy as it moves north through the Caribbean. Use the grid lines *(latitude* and *longitude* lines) on the map to chart Windy's progress. Make a hurricane symbol— —to show the hurricane's location at each point. Then answer the questions that follow. Use the back of this page if you need more space.

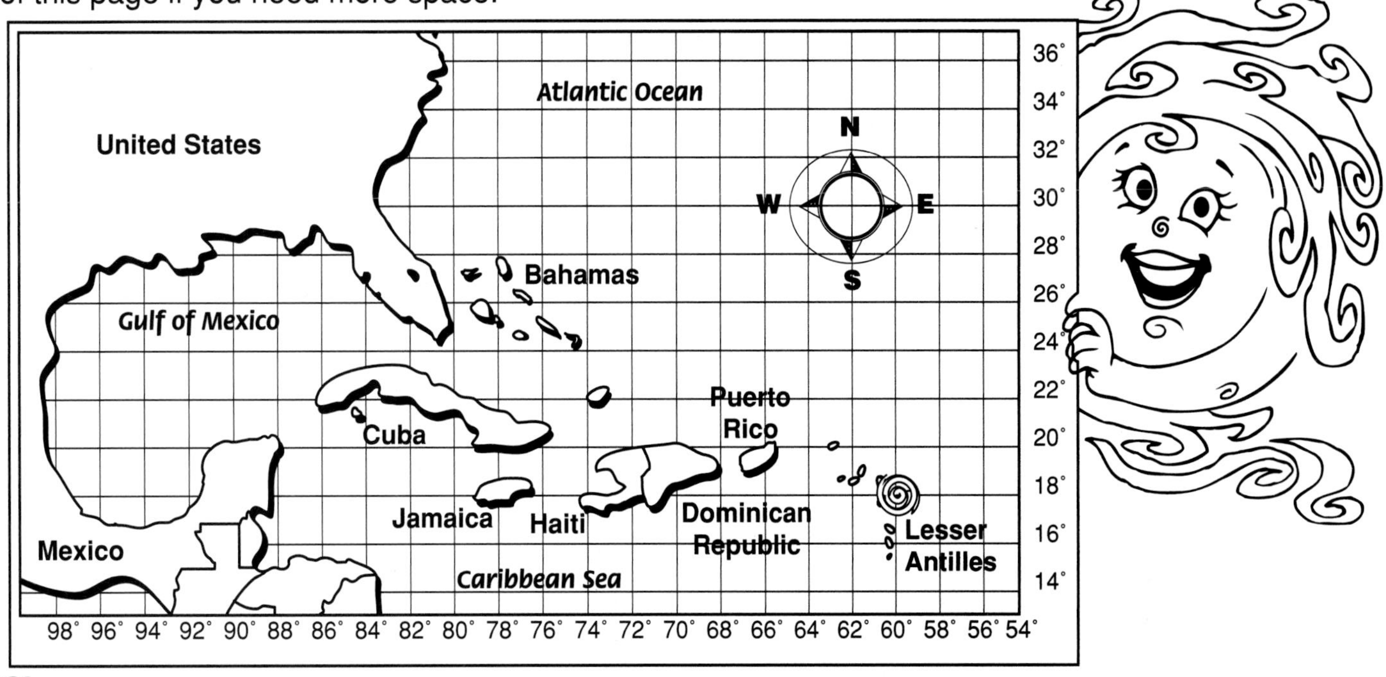

Clues:

1. Hurricane Windy is spotted off the Lesser Antilles. (18°N, 60°W)
2. The *trade winds,* or large wind belts north of the equator, push Windy to the northwest. (21°N, 63°W)
3. Windy continues to move in a westerly direction. (21°N, 67°W)
4. Windy passes near the Dominican Republic and Haiti. (22°N, 70°W)
5. Winds push Windy northwest. (23°N, 75°W)
6. Passing through the Bahamas, Windy heads for Florida. (25°N, 78°W)
7. As Windy approaches the mainland, it continues north and travels along the coast. (28°N, 78°W)
8. Windy moves toward the coasts of Georgia and South Carolina. (31°N, 80°W)
9. The *prevailing westerlies,* strong winds blowing around the earth from west to east, push Windy back out to sea! (33°N, 77°W)
10. North Carolina gets heavy rains, but Windy continues in a northeasterly fashion. (36°N, 75°W)

Questions:

1. Which islands were directly hit by Windy? ______________________________
2. Which island do you think got heavier rains from Windy: Puerto Rico or the Dominican Republic? Why? ______________________________
3. What caused Windy to turn east, back out to sea? ______________________________
4. Do you think the people in Florida should have been evacuated when Windy passed through the Bahamas? Why or why not? ______________________________

Name ______________________ Reading a weather map

Breezin' Through a Weather Map

Use the map shown and its key to answer the questions below.

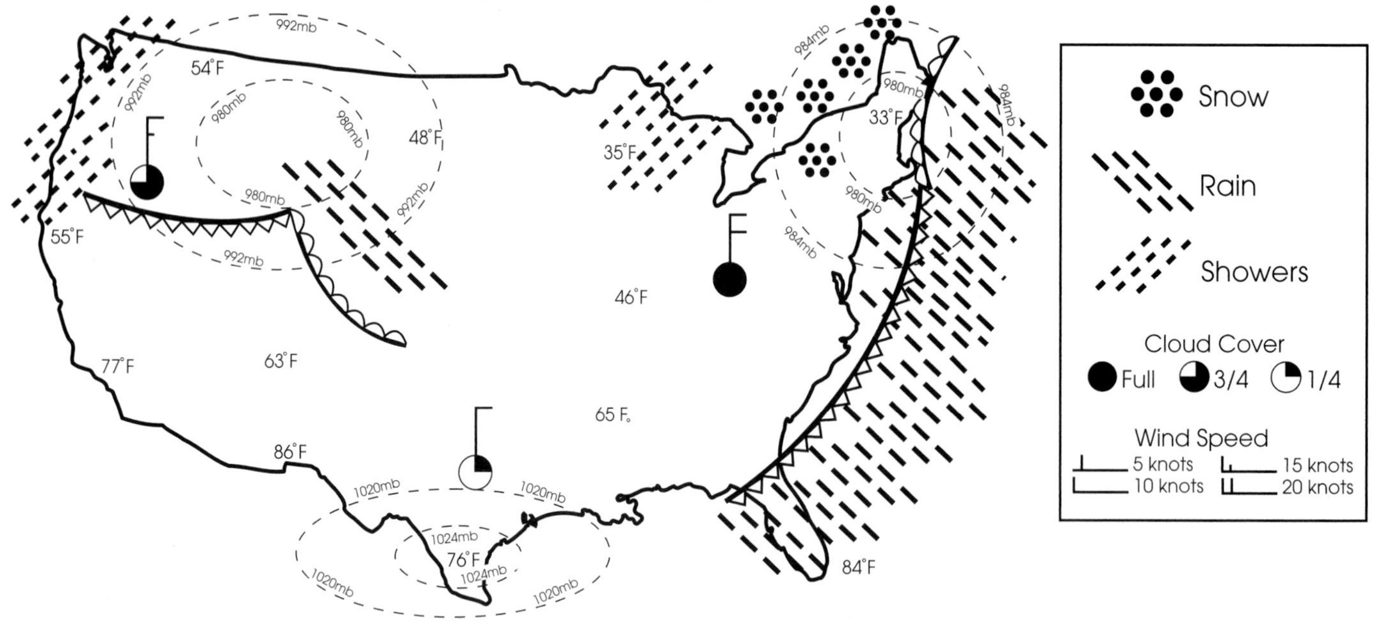

1. Look at the thin dashed lines on the map above. These are *isobars.* They show areas of high and low pressure. Trace the lines with a green pencil.
2. Notice that each isobar line connects air-pressure readings of the same number of *millibars* (units of atmospheric pressure) to create a ring. Compare the numbers on each set of rings, beginning with the outermost ring and moving toward the center. If the numbers decrease, write "LOW" in the centermost ring (an area of low pressure). If the numbers increase, write "HIGH" in the centermost ring (an area of high pressure).
3. A line of connected triangles indicates a cold front, which can bring lower temperatures, wind, rain, or even snow. Color the triangles of each cold front blue.
4. A line of connected semicircles indicates a warm front, which can bring rising temperatures, rain, or snow. Color the semicircles of the warm front red.
5. Look at the large dots that are either fully or partially shaded in. These dots indicate the amount of cloud cover in an area. Is it cloudier in the eastern half of the United States or the western half?

   ____

6. Notice that the large dots have flags attached to them. The flags show the wind speed in each area. Print the wind speed next to each flag. What is the windiest part of the United States: the East, the Northwest, or the South? ______________________
7. What is the highest wind speed recorded on the map? __________ Where are these winds located?

   ____

8. Which parts of the United States are getting rain? ______________________

   Just showers? ______________________
9. Which part of the United States is getting snow? ______________________
10. What is the highest temperature recorded on the map? __________ The lowest? __________

Note to the teacher: Students need green, blue, and red colored pencils to complete this activity.

Solar System Mobile

This project helps students visualize how far each planet is from the sun. Make sure students understand that this model shows the relative distance from each planet to the sun. Let them know it does not show the distance of the planets from each other or the accurate size of each planet.

Materials for each student:

copies of pages 111 and 112
3" x 20" strip of yellow tagboard
string or yarn
ruler
access to a hole puncher
tape

Day 1: Have students complete page 111. Discuss each planet's distance from the sun with students.

Day 2: After reviewing each planet's distance from the sun, guide students through the steps below.

Steps for students:

1. Label the tagboard strip "Sun."
2. About one-half inch from the bottom of the tagboard strip, use a ruler to make eight marks two inches apart.
3. Punch one hole for each mark on the tagboard.
4. Cut out the patterns on page 112.
5. If time allows, color each planet cutout so it resembles the actual planet.

Day 3: Have students complete the steps below to finish their projects.

Steps for students:

1. Tie one end of the string to the first hole in the tagboard. Refer to column 4 on page 111 and measure and cut the string to show the distance from Mercury to the sun.
2. Tape the circle labeled "Mercury" to the end of the string, getting the string as close as possible to the end of the circle.
3. Repeat Steps 1 and 2 with each of the other planets until all eight planets are hanging in order.
4. Tape the ends of the tagboard together to create a loop that represents the sun. Then punch a hole at the top of two sides of the sun and attach a length of string to make a hanger.

All in a Day

Materials:
chart with the information shown
12" x 18" sheet of construction paper for each student

Explain to students that the 24 hours it takes the earth to make one rotation (complete turn on its axis) equals one Earth day. Display the chart with the rotation times of each planet. Have volunteers share how the difference in the length of a day would change their lives. Then instruct each child to cut a circle from her construction paper. Next, have each student imagine she lives on a planet other than Earth. Have her write in a spiral shape on one side of her paper circle explaining how she would spend a day on her chosen planet. (Remind each child to make sure her activities add up to one complete day on her chosen planet.) Have her draw illustrations of her day on the other side of her circle.

Planet	One Rotation
Mercury	about 59 Earth days
Venus	about 243 Earth days
Earth	about 24 Earth hours
Mars	about 24.5 Earth hours
Jupiter	about 10 Earth hours
Saturn	about 10.5 Earth hours
Uranus	about 17 Earth hours
Neptune	about 18.5 Earth hours

Planetary Tilts

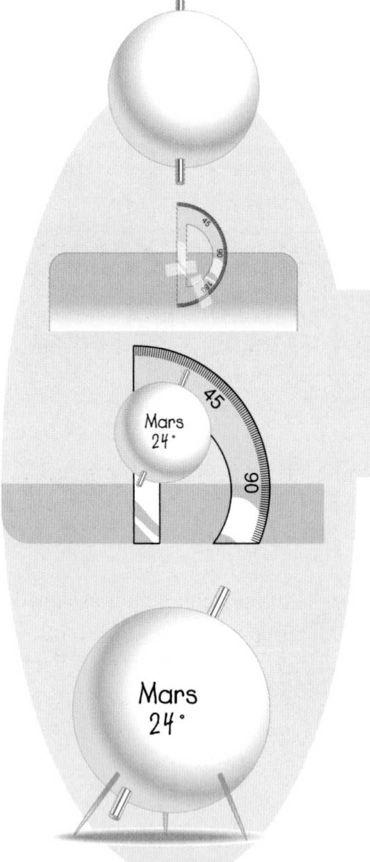

Help students visualize the differences in the tilts of the planets by making these simple models. Explain to students that each planet has an imaginary axis through its center, tilted at an angle. Display a chart of planetary tilts, similar to the one shown. Then divide students into eight groups and assign each group a different planet. Guide each group through the steps below to make a model of its planet. Display the finished models so students can examine the tilts of each planet.

Materials for each group of students:
sharpened pencil
4" length of a plastic drinking straw
3" Styrofoam ball
permanent marker
ruler
tape
3 toothpicks
protractor

Steps for students:
1. Place the Styrofoam ball (representing your planet) on the table or desk. Use a sharpened pencil to carefully make a hole through the center of the ball. Then push the straw (axis) through the hole until it is sticking out on both ends.
2. Label the ball with the planet's name and the correct number of tilt degrees (the angle at which a planet's axis is tilted) from the chart.
3. Tape the protractor to a desk so that the 90° mark lines up with the desktop as shown.
4. Hold the planet up to the protractor and line up the axis to the correct number of tilt degrees. Stick three toothpicks in the bottom of the ball so that it stands at the right number of tilt degrees. (You may have to rearrange the toothpicks until the ball stands at the correct number of tilt degrees for your planet.)

Planet	Mercury	Venus	Earth	Mars	Jupiter	Saturn	Uranus	Neptune
Planetary Tilt	near 0°	178°	23½°	24°	3°	27°	98°	29°

Planetary Weather Reports

Students will learn about the weather on other planets with this role-playing activity.

Materials for each group of students:
access to reference materials

Day 1: Have students discuss what types of information are usually presented in a weather report, such as temperatures, cloud cover, precipitation levels, and wind speeds. Then divide students into seven groups and assign each group a planet other than Earth. Have each group begin to research weather facts for its assigned planet.

Day 2: Have each group complete its research on weather facts for its planet. Then instruct each group to write a short weather report describing the weather on its planet. When the reports are complete, have a weather reporter from each group read the weather report for his group's planet.

Moon Explorations

Chronicle humanity's attempts to learn about the moon with this group research activity.

Materials for each group of students:
access to research materials
blank card

Day 1: Divide students into eight groups and assign each group a different event from the chart shown. Have each group begin to research its assigned event.

Day 2: Have students complete research on their assigned lunar event. Then have each group write the year(s) its event took place on a blank card. Direct the groups to stand in chronological order. Then have each group, in turn, tell about its event. If time permits, lead students in discussing what future moon explorations might be like.

1959	*Luna 2* and *Luna 3* (Soviet Union)
1964–1965	*Ranger VII, Ranger VIII, Ranger IX* (United States)
1966	*Luna 9* (Soviet Union)
1968	*Apollo 8* (United States)
1969	*Apollo 11* and *Apollo 12* (United States)
1970	*Luna 16* (Soviet Union) and *Apollo 13* (United States)
1971	*Apollo 14* and *Apollo 15* (United States)
1972	*Apollo 16* and *Apollo 17* (United States)

The Phases of the Moon

Help students understand the phases of the moon with this demonstration. As each moon phase is discussed, direct each child to draw and label a picture of the moon phase on the corresponding page in his booklet. When the booklets are complete, invite each child to attach a construction paper cover and add a title.

Materials:
large, dark-colored ball, such as a basketball or kick ball, with an X drawn on one side (to remind students that the same side of the moon always faces the earth)
large flashlight
8-page booklet for each child
sheet of construction paper for each child

Steps:

1. Have one child hold the flashlight (sun). Have another child hold the ball (moon) up and away from his body so that the X is facing him and he is looking up a little to see the ball. Have the student with the flashlight stand a few feet in front of the child holding the ball, and shine the light on the student holding the ball.
2. Dim the lights and point out that the moon is dark on the side that the child holding it sees. Tell students this is a new moon.
3. Have the student holding the ball slowly turn counterclockwise until he sees a small sliver of light on the ball. Tell students this is a waxing crescent moon.
4. Instruct the student to continue turning until about half of his side of the ball is illuminated. Explain to students this is a first-quarter moon.
5. Have the student continue turning until only a small sliver of his side of the ball is not illuminated. Tell students this is a waxing gibbous moon.
6. Direct the student to continue turning until all of the side he can see is illuminated. Tell students this is a full moon.
7. Have the student keep turning until only a small sliver of his side of the ball is not illuminated. Tell students this is a waning gibbous moon.
8. Have the student keep turning. Have him stop when about half of his side of the ball is dark. Inform students this is the third-quarter moon.
9. Instruct the student to keep turning until he again sees only a small sliver of light on the ball. Tell students this is a waning crescent moon.
10. Ask students if the ball decreased in size as the student rotated. (No.) Then have your students explain what did change that made the moon appear to change shape. *(The amount of visible sunlight hitting the moon changed, making it only appear to change shape.)*

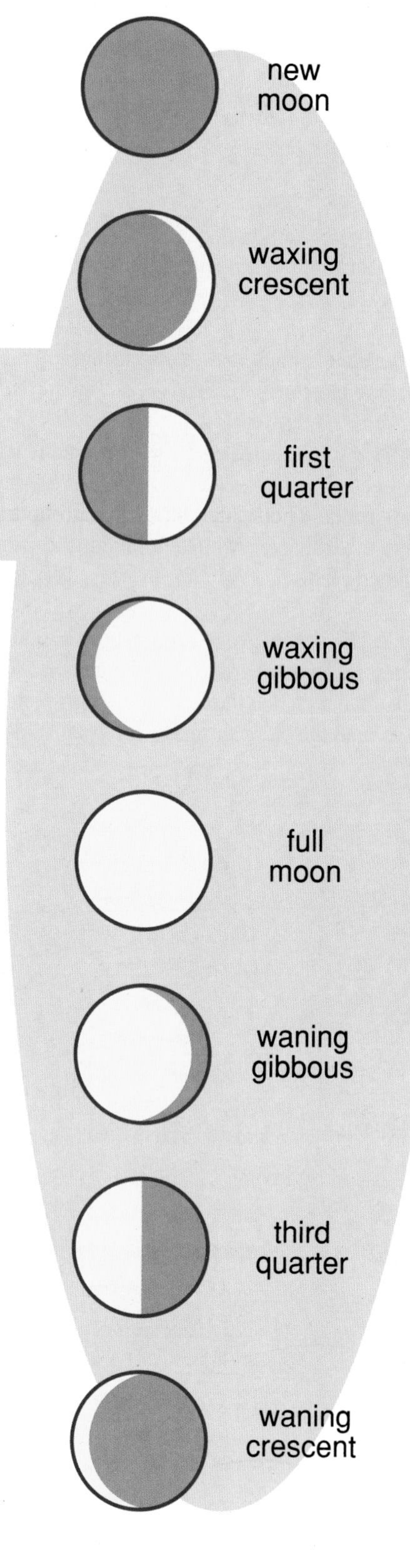

Did You Know?
A waxing moon is one that appears to grow larger, while a waning one appears to grow smaller.

Earth & Space Science

Time Will Tell!

On a sunny morning, have students experiment with an ancient method of telling time. Divide students into small groups and guide them through the steps below to make a sundial.

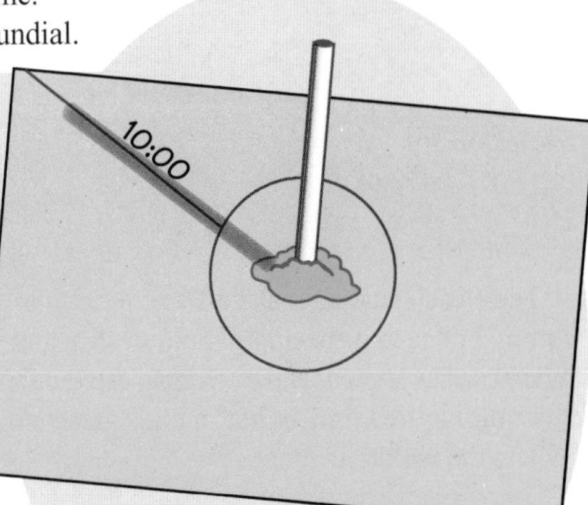

Materials for each group of students:
12" x 18" construction paper
lump of clay
5" length of straw
ruler

Steps for students:
1. Place the clay in the center of the paper. Place the straw upright in the clay.
2. Draw a circle with a five-inch radius around the base of the clay, using the clay as the center of the circle.
3. Place the paper in a sunny location.

Follow-up: One hour later, have each group trace the straw's shadow on the paper. If necessary, direct them to extend the tracing until it touches the circle's edge. Have each group label this point with the hour of the day. Instruct the groups to repeat this process each hour throughout the school day. At the end of the day, discuss with students what they observed and why. *(The earth's rotation makes the sun appear to move in the sky each day. It takes the earth 24 hours to make one complete rotation. Students can measure the time by dividing the sundial into 12 equal parts for the daylight hours.)*

Moving Sunspots?

Tell students that sunspots are dark, cooler areas of the sun's surface. Also share that most sunspots are outlined by a lighter-colored area called the penumbra; they follow a path parallel to the solar equator (which makes it easy to observe the east-to-west rotation of the sun), and they appear to move due to the sun's rotation. Then guide students through the steps below to make a model that illustrates this phenomenon.

Sunspots
Sun
outer, lighter part (penumbra)
dark central part (umbra)
March 6
March 10
March 14
1 2
1 2
2
Movement of sunspots due to the Sun's east-to-west rotation

Materials for each student:
sheet of construction paper
yellow paper circles—1 large and 3 small
black crayon

Steps for students:
1. On the large paper circle, use the crayon to draw a dark, irregular spot in two different places.
2. Around each dark spot, draw a fuzzy outer band that is light in color.
3. Label the large circle as shown. Then glue it to the left side of the construction paper.
4. On the right side of the paper, glue the smaller circles, overlapping them slightly.
5. Beginning with the circle closest to the largest circle, draw on each small circle (in a path parallel to the sun's equator) and label, in order, the position of the sunspots as they would appear several days later due to the sun's rotation.

Did You Know?
Sunspots usually appear in 11-year cycles, reaching the greatest number every 11 years. Sunspots can vary in size and intensity.

Star Vocabulary

Materials for each student:
copy of the top of page 113
15 4" x 6" index cards

UNIVERSE

Everything that we know exists and believe may exist

Have each student make a set of vocabulary review flash cards by cutting out each word-and-definition strip and wrapping it around an index card as shown. Have students use the flash cards to quiz each other during free time or play a class game similar to Pictionary to review the words.

Did You Know?
A star shines due to a nuclear reaction between hydrogen atoms in its core, causing the release of light energy.

Twinkling Stars

This simple experiment helps students understand why stars twinkle.

Materials:
medium-size glass bowl two-thirds full of water
piece of tagboard sized to fit under the bowl
aluminum foil
flashlight

Steps:
1. Tear the foil into small pieces and shape the pieces into small balls so they resemble stars.
2. Place the foil pieces on the tagboard.
3. Set the glass bowl atop the tagboard and dim the lights.
4. Shine the flashlight on the bowl and have students observe the stars.
5. While still shining the flashlight on the bowl, tap the bowl, causing the water to move.
6. Have students describe what the stars do when the water moves. *(They appear to twinkle.)*
7. Explain to students that the stars appear to twinkle because, when the water is disturbed, the light rays are refracted, or bent at different angles, just as the light from the stars is refracted at different angles by the difference in air densities in the earth's atmosphere.

Into the Black Hole

Materials:
large bedsheet
marble
stack of textbooks

Help students see how objects get sucked into a black hole with this simple demonstration. Explain to students that a black hole is the result of a collapsing star. Further explain that a black hole has such a strong gravitational force that not even light can escape from it. Then place the bedsheet on the ground in an open area and stack several textbooks in the center of the sheet. Have students stand around the sheet, pick up the edges, and hold it taut at waist level. Direct several student volunteers, in turn, to try to roll the marble from one side of the sheet to the other. Then have students discuss the results of the activity and how they are similar to the effects of an actual black hole in space. *(The marble will travel in a curved path toward the textbooks, just as gravity pulls space objects into a black hole.)*

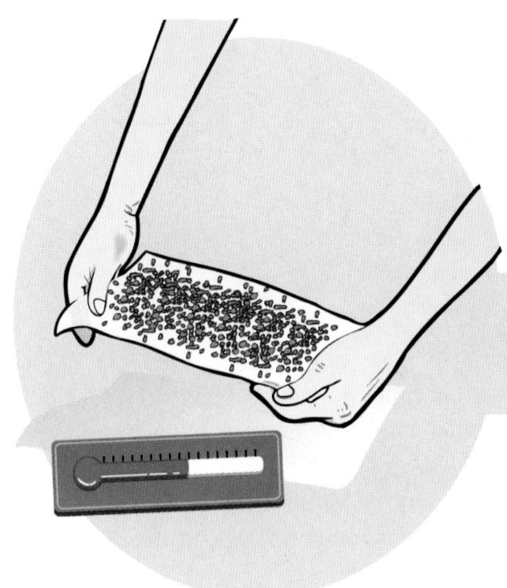

The Big Chill

Use this experiment to show students how a large meteorite striking the earth might temporarily change its climate. Divide students into groups of three and guide them through the steps shown, calling out the times at which they need to record the temperature and complete the next step.

Materials for each group of students:
thermometer
piece of plastic wrap
dry dirt

Steps for students:

1. Place the thermometer on the ground outside, in direct sunlight. Wait four minutes; then record the temperature.
2. Hold a piece of plastic wrap about eight inches above the thermometer and sprinkle some dry dirt (dust) on the plastic wrap. Make sure the thermometer is in the shadow cast by the plastic wrap and the dust. Wait two minutes; then record the temperature.
3. Without disturbing the thermometer, add more dry dirt to the plastic wrap. Wait two minutes; then record the temperature again.

Follow-up: Discuss the results with the class. Have students share ways a lowered temperature on Earth might affect life on the planet. *(A large meteorite would send tremendous amounts of dust into the atmosphere, effectively blocking the sun. Without sunlight, the temperature on Earth would quickly drop.)*

Deep Impact

Examine the formation of impact craters made by meteorites with a demonstration that uses different-size potatoes. Tell students that impact craters are bowl-shaped depressions measuring up to about ten miles in diameter.

Materials:
3–4 sheets of paper
10" pie plate
flour
Kool-Aid drink mix
3–4 different-size potatoes, ranging in size from small to medium
meterstick
old newspapers

Steps:
1. Fill the pie plate with one inch of flour; then cover the surface with a thin layer of Kool-Aid drink mix (so the results can be more easily observed). Then place the pie plate atop the spread-out newspapers.
2. From a height of about one meter, drop a potato (meteorite) into the plate.
3. Carefully remove the potato; then have student volunteers help measure and record the diameter of the impact crater made by your meteorite. Also ask the volunteers to measure and record the depth of the impact crater and how far the flour from the crater was thrown.
4. Repeat Steps 1–3 with each remaining potato. Then lead students in discussing the results and comparing the effects of the different-size meteorites. *(The greater the size of the meteorite, the greater the impact crater and the area of debris. Scientists think that a large meteorite crashing into the earth would send huge amounts of dust into our atmosphere, dangerously cooling our planet.)*

Modern-Day Myths

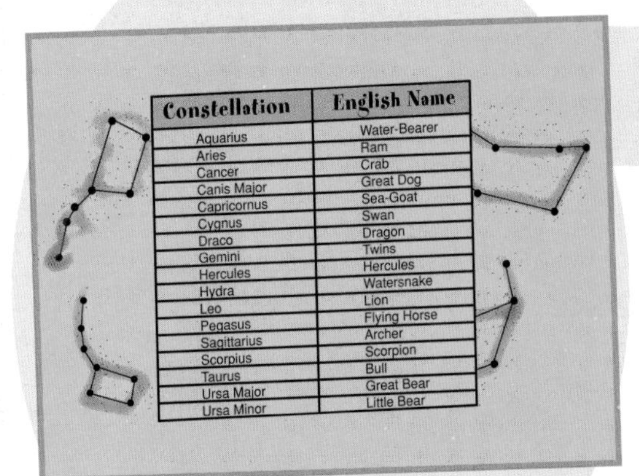

Constellation	English Name
Aquarius	Water-Bearer
Aries	Ram
Cancer	Crab
Canis Major	Great Dog
Capricornus	Sea-Goat
Cygnus	Swan
Draco	Dragon
Gemini	Twins
Hercules	Hercules
Hydra	Watersnake
Leo	Lion
Pegasus	Flying Horse
Sagittarius	Archer
Scorpius	Scorpion
Taurus	Bull
Ursa Major	Great Bear
Ursa Minor	Little Bear

Materials for each student:
copy of the bottom of page 113
access to reference materials and maps of constellations

Day 1: Explain to students that in ancient times people made up stories, called myths, to explain things they didn't understand. Further explain that many constellations are named for mythological characters such as gods, goddesses, and heroes. Direct each child to choose a constellation from her copy of page 113. Have her research her constellation to find one or more constellations that are nearest in location to the one she has chosen.

Days 2 and 3: Have each child write a brief modern-day myth using her chosen constellation as the main character and the nearest constellation(s) as the supporting character(s). When each student is finished, invite her to share her myth with the class.

See the space skill sheets on pages 114–121.

Name____

To Mercury and Beyond

Compare the distance from each planet to the sun by completing the chart below. The chart shows the distance from each planet to the sun in kilometers. Follow the directions to create a new scale that shows the distances in more manageable numbers.

Directions:

1. Divide Earth's distance from the sun by 10 million. (Dividing by 10 million is the same as moving the decimal seven places to the left.) This will convert the distance to a more manageable scale (14.9600000). Write this number in column 3.
2. Round this number to the nearest whole number; then write the number in column 4 (14.9600000 = 15).
3. Repeat Steps 1 and 2 for the other seven planets.

1 Planet	2 Distance From the Sun (in kilometers)	3 New Scale (10 million kilometers = 1 millimeter)	4 New Scale Rounded to Nearest Whole Number (in millimeters)
Mercury	57,900,000.0		
Venus	108,200,000.0		
Earth	149,600,000.0	14.9600000	15
Mars	227,900,000.0		
Jupiter	778,400,000.0		
Saturn	1,429,400,000.0		
Uranus	2,875,000,000.0		
Neptune	4,504,300,000.0		

Bonus: Choose three of the numbers from column 2 and write them in word form.

Note to the teacher: Use with "Solar System Mobile" on page 103.

Planet Patterns

Use with "Solar System Mobile" on page 103.

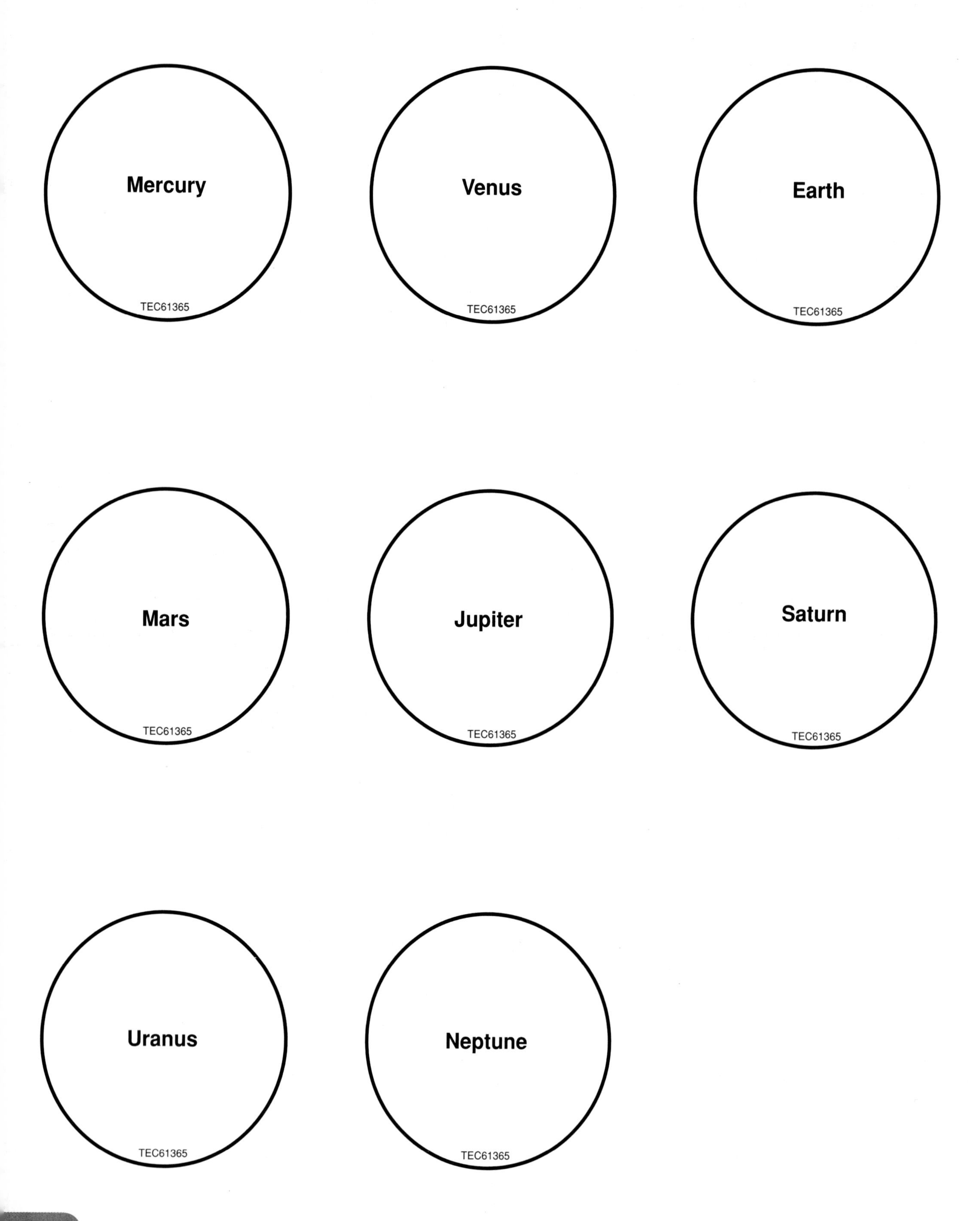

Use with "Star Vocabulary" on page 108.

ASTRONOMY	Study of the planets, stars, and other objects in the universe TEC61365
BLACK HOLE	Collapsed stars and other objects whose gravitation is so strong that not even light rays can escape TEC61365
CONSTELLATION	Groups of stars that can be seen in the night sky TEC61365
GALAXY	The largest star cluster TEC61365
GRAVITY	Force that pulls objects together because of their mass TEC61365
NOVA	Exploding star that becomes thousands of times brighter than normal before becoming dim again TEC61365
ORBIT	In space, the path one body takes around another TEC61365
SOLAR SYSTEM	The sun and all the planets and other bodies that orbit it TEC61365
STAR	Enormous ball of glowing gases TEC61365
STAR CLUSTERS	Groups of hundreds to millions of stars TEC61365
SUPERNOVA	Exploding star that is thousands of times brighter than an ordinary nova TEC61365
TELESCOPE	Tool used to see stars in our galaxy and beyond TEC61365
UNIVERSE	Everything that we know exists and believe may exist TEC61365
WEIGHT	Force of gravity placed on an object TEC61365
WHITE DWARF	Star of the second smallest known type TEC61365

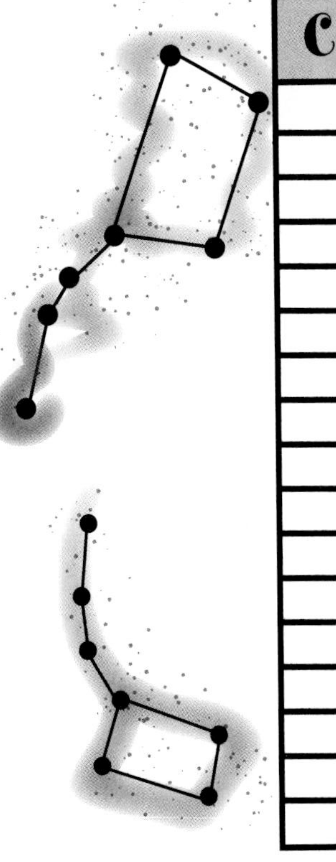

Constellation	English Name
Aquarius	Water-Bearer
Aries	Ram
Cancer	Crab
Canis Major	Great Dog
Capricornus	Sea-Goat
Cygnus	Swan
Draco	Dragon
Gemini	Twins
Hercules	Hercules
Hydra	Watersnake
Leo	Lion
Pegasus	Flying Horse
Sagittarius	Archer
Scorpius	Scorpion
Taurus	Bull
Ursa Major	Great Bear
Ursa Minor	Little Bear

Note to the teacher: Use with "Modern-Day Myths" on page 110.

Name______________________

Planetary Contract

Have fun completing these far-out planetary activities!
Color each planet after completing its activity.

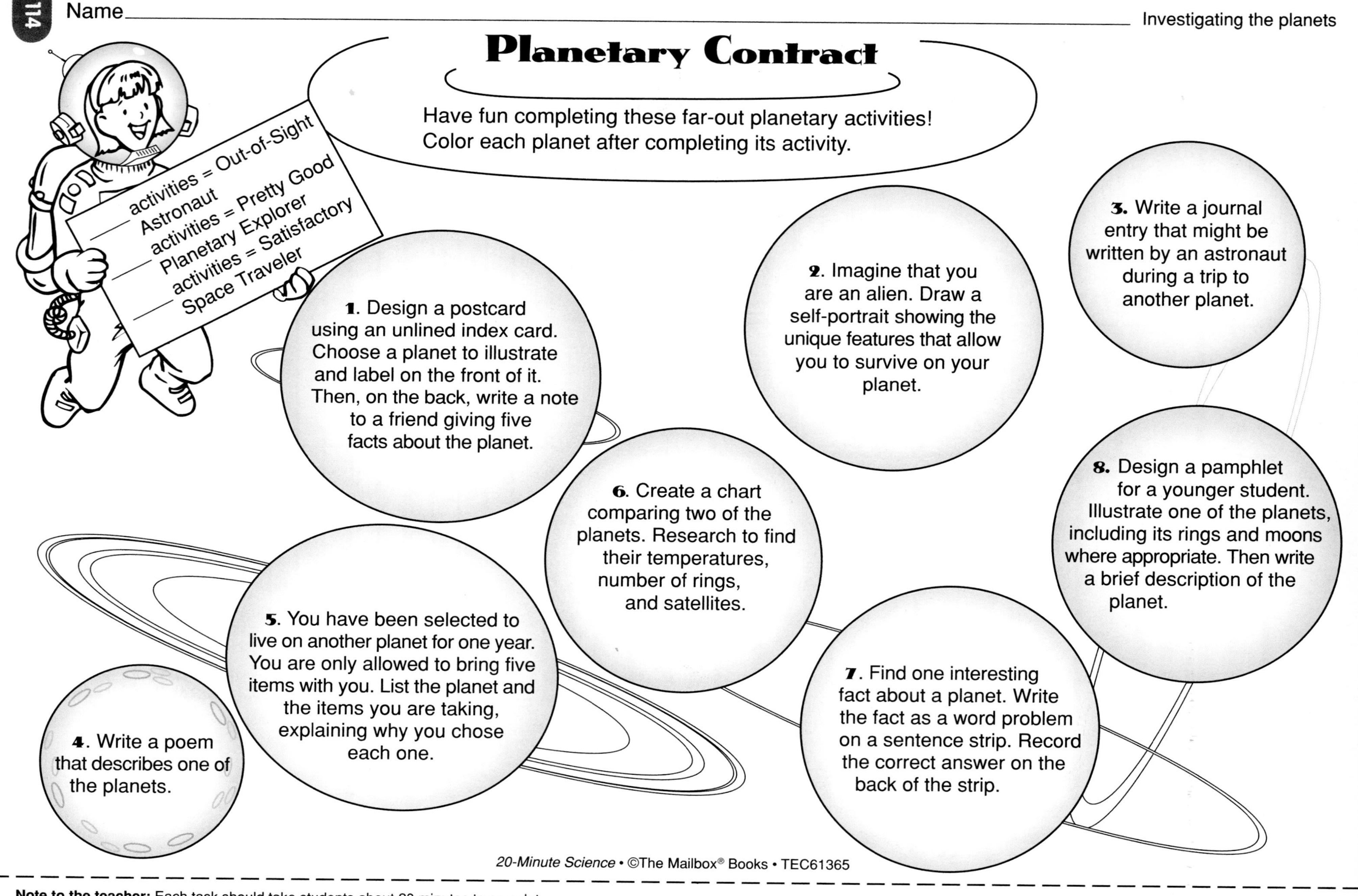

Note to the teacher: Each task should take students about 20 minutes to complete.

Name ______________________________

Wait, That's Space Weight!

A planet's gravitational pull, or pulling force, keeps everything on the planet from flying off into space. Each planet's gravitational pull is different because each planet's mass is different. Your weight on a planet shows how much gravity is pulling you down to the planet's surface.

I. Orbit, the space dog, wants to know how much he would weigh if he traveled to each planet. Multiply each planet's surface gravity **(SG)** by Orbit's Earth weight (25 pounds) to help him find out how much he would weigh. Fill in the space helmets with the correct answers.

1. Mercury SG = .38 ________
2. Venus SG = .91 ________
3. Earth SG = 1 ________
4. Mars SG = .38 ________
5. Jupiter SG = 2.54 ________
6. Saturn SG = 1.07 ________
7. Uranus SG = .90 ________
8. Neptune SG = 1.15 ________

ORBIT

II. To find out how much you would weigh if you traveled to each of the planets, multiply each planet's surface gravity by your weight. Fill in the space boots with the correct answers.

Mercury ________
Venus ________
Earth ________
Mars ________
Jupiter ________
Saturn ________
Uranus ________
Neptune ________

Planet Power

Read each question and write your answers in the spaces provided. Next, look at the alien in each box and decide if you have earned the number of points it is holding for answering each question correctly. Then color one band of the Planet Power Meter for every point you feel you have earned. After answering all the questions, check your Planet Power rating!

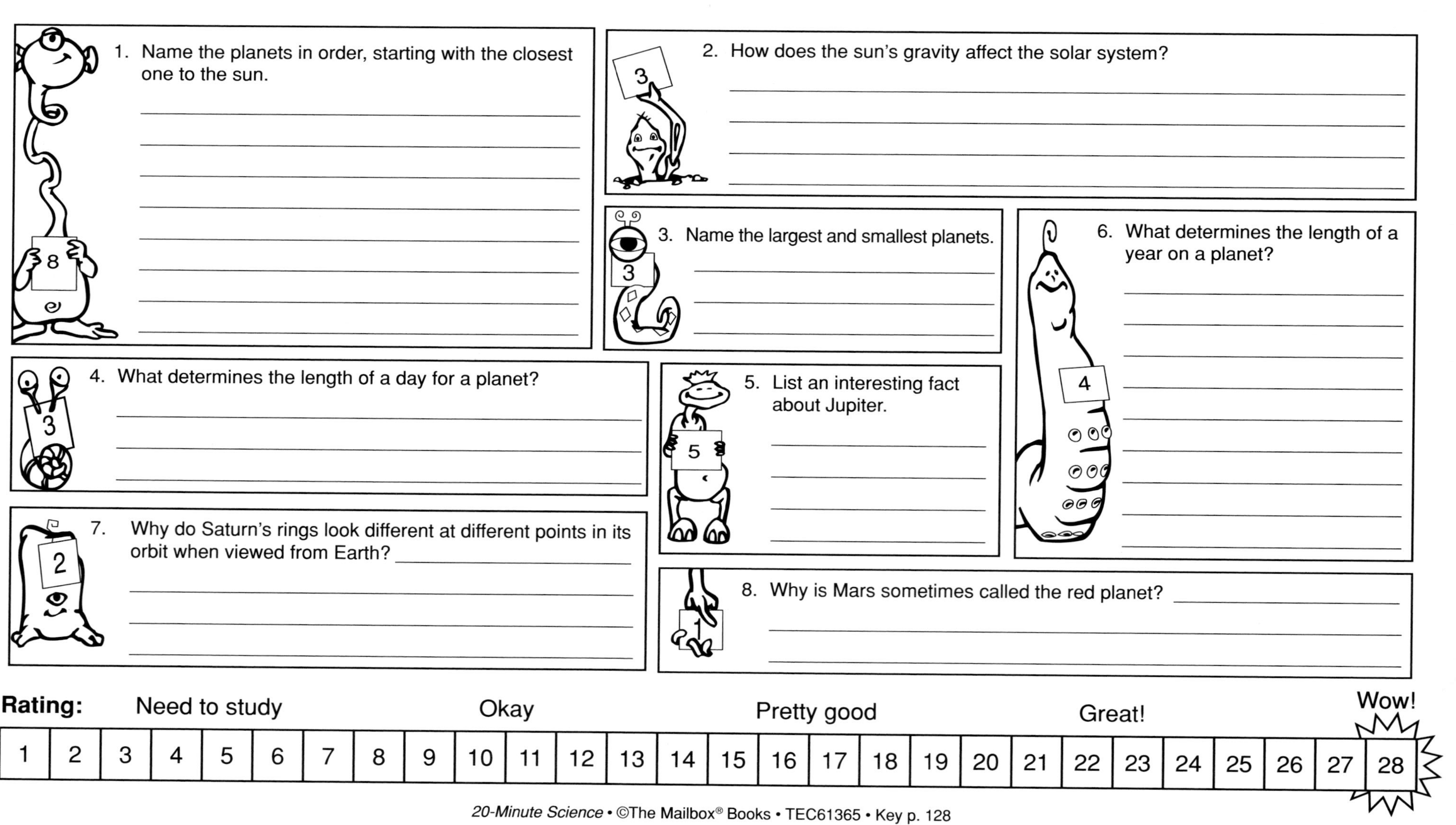

1. Name the planets in order, starting with the closest one to the sun.
2. How does the sun's gravity affect the solar system?
3. Name the largest and smallest planets.
4. What determines the length of a day for a planet?
5. List an interesting fact about Jupiter.
6. What determines the length of a year on a planet?
7. Why do Saturn's rings look different at different points in its orbit when viewed from Earth?
8. Why is Mars sometimes called the red planet?

Rating: Need to study — Okay — Pretty good — Great! — Wow!

1	2	3	4	5	6	7	8	9	10	11	12	13	14	15	16	17	18	19	20	21	22	23	24	25	26	27	28

Are We There Yet?

Feel like planning a trip? How about traveling to Mars or even Neptune? Use the mileage chart shown below to help you answer the following questions. Complete each problem on the back of this page; then write your answer on the line provided.

	Sun	**Mercury**	**Venus**	**Earth**	**Mars**	**Jupiter**	**Saturn**	**Uranus**	**Neptune**
Sun	0	36	67	93	141	480	900	1,800	2,800
Mercury	36	0	31	57	105	444	864	1,764	2,764
Venus	67	31	0	26	74	413	833	1,733	2,733
Earth	93	57	26	0	48	387	807	1,707	2,707
Mars	141	105	74	48	0	339	759	1,659	2,659
Jupiter	480	444	413	387	339	0	420	1,320	2,320
Saturn	900	864	833	807	759	420	0	900	1,900
Uranus	1,800	1,764	1,733	1,707	1,659	1,320	900	0	1,000
Neptune	2,800	2,764	2,733	2,707	2,659	2,320	1,900	1,000	0

Mileage in Millions

1. How far is Jupiter from Earth? ____________________
2. How far is Saturn from Uranus? ____________________
3. Which would take longer: traveling from Earth to Mars or from Mercury to Venus?
   ____
4. Which would take less time: traveling from Saturn to Jupiter or from Jupiter to Venus?
   ____
5. How much farther is Mercury from Uranus than it is from Earth? ____________________
6. If you traveled from Neptune to Mars and then from Mars to Venus, how many miles would you travel? ____________________
7. Which planet is closer to Mercury: Jupiter or Saturn? ____________________ How many miles closer? ____________________
8. How much closer is Saturn to Uranus than Uranus is to Neptune? ____________________

Name________________________________

Going Moon Gazing

Observe the moon each night for two weeks. Record the date and time of each observation in the chart below. Include a hand-drawn picture of how the moon looks.

Note: If the sky is cloudy, making it difficult to observe the moon, try again later. If it cannot be observed at all, write "not visible" in the appearance box.

Week 1 | **Week 2**

Date	Time	Appearance	Date	Time	Appearance

Moon Phases

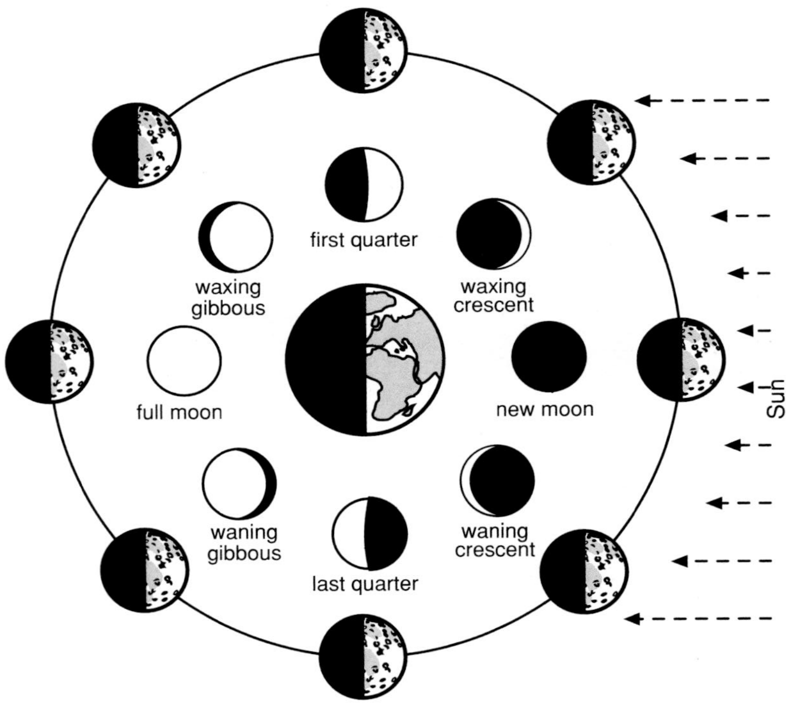

first quarter
waxing gibbous
waxing crescent
full moon
new moon
waning gibbous
waning crescent
last quarter
Sun

1. What did you notice about the appearance of the moon during the first seven days? During the last seven days?

____

____

2. What do you predict the moon will look like during the last week before the next new moon?

____

____

Note to the teacher: Students will need 20 minutes or less to complete each night's observation.

Name ____

Moon Minibook

Write a word from the bottom of the page in each blank in the boxes below. Use each answer only once. Then cut apart the pages and staple them on the left side to make a minibook. Use reference materials to help you.

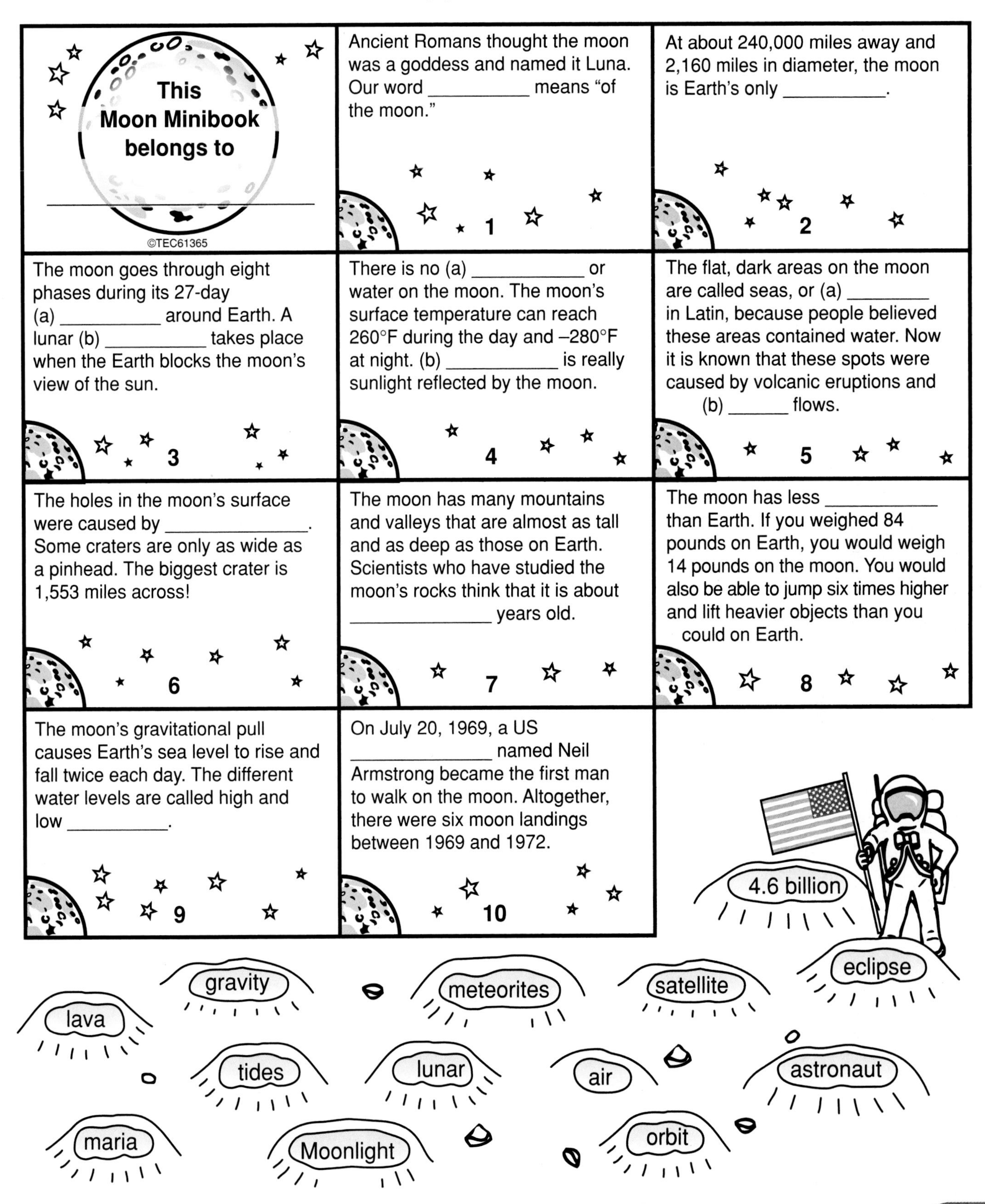

Ancient Romans thought the moon was a goddess and named it Luna. Our word __________ means "of the moon."

1

At about 240,000 miles away and 2,160 miles in diameter, the moon is Earth's only __________.

2

The moon goes through eight phases during its 27-day (a) __________ around Earth. A lunar (b) __________ takes place when the Earth blocks the moon's view of the sun.

3

There is no (a) __________ or water on the moon. The moon's surface temperature can reach 260°F during the day and –280°F at night. (b) __________ is really sunlight reflected by the moon.

4

The flat, dark areas on the moon are called seas, or (a) ________ in Latin, because people believed these areas contained water. Now it is known that these spots were caused by volcanic eruptions and (b) ______ flows.

5

The holes in the moon's surface were caused by ______________. Some craters are only as wide as a pinhead. The biggest crater is 1,553 miles across!

6

The moon has many mountains and valleys that are almost as tall and as deep as those on Earth. Scientists who have studied the moon's rocks think that it is about ______________ years old.

7

The moon has less ___________ than Earth. If you weighed 84 pounds on Earth, you would weigh 14 pounds on the moon. You would also be able to jump six times higher and lift heavier objects than you could on Earth.

8

The moon's gravitational pull causes Earth's sea level to rise and fall twice each day. The different water levels are called high and low __________.

9

On July 20, 1969, a US ______________ named Neil Armstrong became the first man to walk on the moon. Altogether, there were six moon landings between 1969 and 1972.

10

Name ______________________________

Dirty Snowballs

A comet can sometimes be seen with the naked eye, appearing as a fuzzy star in the night sky as it orbits the sun. Sometimes its nucleus is described as resembling a dirty snowball because it's made up of various kinds of ice and rocky particles. A comet usually has one or two tails, made up of dust, gas, or both. The table below gives information about the first sightings of some comets and their orbits. Use the information to answer the questions that follow.

Name of Comet	First Recorded Sighting	Average Orbital Period in Years
Halley's Comet	about 240 BC	76
Comet Swift-Tuttle	69 BC	130
Comet Tempel-Tuttle	1366	33
Biela's Comet	1772	6.6
Encke's Comet	1786	3.3
Comet Flaugergues	1811	3,100
Great Comet	1843	513
Great September Comet	1882	759
Comet Schwassmann-Wachmann 1	1927	15
Comet Ikeya-Seki	1965	880
Comet Bennett	1969	1,678
Comet West	1975	558,300

1. Which comets are you not likely to see in your lifetime? ______________________________

____

2. Halley's Comet last crossed Earth's orbit in 1986. In what year will it make its next appearance? ______________________________

3. Write the years in which the comets below are next expected to pass by Earth.
   a. Comet Ikeya-Seki ______________
   b. Comet Schwassmann-Wachmann 1 ______________
   c. Comet Bennett ______________
   d. Comet Flaugergues ______________
   e. Great Comet ______________

4. Short-period comets take less than 200 years to orbit the sun; long-period comets take more than 200 years. In the above table, color the short-period comets yellow and the long-period comets blue.

Name ______________________________ Asteroid facts

Asteroids in Agreement

Each sentence below is a fascinating fact about asteroids and contains a subject and a verb that may or may not agree with one another. If the subject in a sentence is singular, it must have a *singular verb*. If the subject in a sentence is plural, it must have a *plural verb*. That way, there is subject-verb agreement.
Note: A singular verb usually ends with an s.

Read each sentence below. Circle the subject and underline its verb in each sentence. If they agree, color the first asteroid at the end of the sentence. If they disagree, color the second asteroid.

	Agree	Disagree
1. Asteroids is chunks of rock that circle the sun.	A	R
2. Ceres are the biggest asteroid ever discovered.	C	E
3. Many asteroids look a lot like giant potatoes.	I	Y
4. Some asteroids' paths brings them near Earth.	L	S
5. Scientists study the reflected light of asteroids to discover what asteroids are made of.	M	U
6. Asteroids vary in size.	U	M
7. Close-up photographs were first taken of an asteroid by the space probe *Galileo.*	N	T
8. Minor planets and planetoids is other names for asteroids.	U	D
9. A Hirayama family are made up of asteroids that travel in the same orbit.	B	A
10. Ceres makes up about one-third of the material in the asteroid belt.	O	F
11. People has never traveled to the asteroid belt.	B	T
12. Most asteroids are found in the asteroid belt.	X	C
13. Astronomers uses radio telescopes to create pictures of asteroids.	Q	G

Solve this mystery sentence by writing each colored letter above on its matching numbered line or lines.

__ __ __ __ __ __ __ __ __   __ __ __ __ __ __ __   __ __ __ __ __ __ __ __ __ __ __ !
9 4 11 2 1 10 3 8 4   9 4 11 10 6 7 8   9 4 11 1 10 7 10 5 2 1 4

Bonus: On the back of this sheet, write two sentences about why you think it's fun studying asteroids. Be sure that your subjects and verbs agree.

Answer Keys

Page 11

1. F
2. E
3. A
4. I
5. C
6. J
7. B
8. H
9. D
10. G

Page 13

Graphs will vary.

Answers for 1, 2, and 6 will vary.

3. When the average child is asleep, her heart beats about 50–70 times per minute.
4. A heart beats slowest during periods of rest and fastest during exercise.
5. The heart would beat faster in order to pump oxygenated blood to the muscles.

Page 14

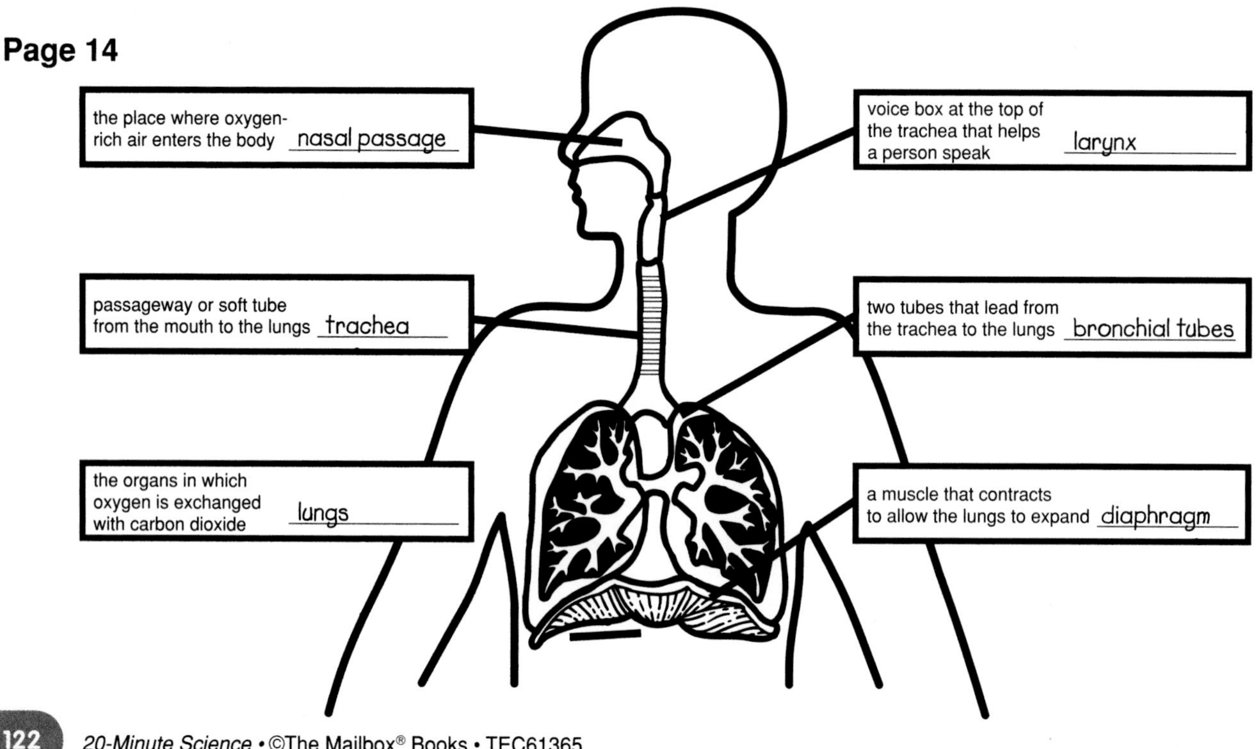

Page 15

Part I: Answers will vary.

Part II: Graphs will vary.

Part III

1. The breathing rate increased with the increase in activity.
2. The breathing rate decreased with the decrease in activity.
3. As the body becomes more active, it uses more oxygen. In order to maintain normal levels of oxygen in the body, the activity of the respiratory system increases to keep up with the body's activity.

Page 16

1. hiccupping, S
2. smiling, C
3. getting goose bumps, Z
4. chewing, R
5. coughing, M
6. adjusted your pupils, O
7. sneezing, L
8. singing, H
9. crying, B

YAWNING

Page 18

When a finger is pricked, a sensory receptor in the skin sends a nerve impulse along an afferent neuron, which is made of an axon, or nerve fiber, that is an extension of the nerve cell. Most axons are less than one millimeter long. Some axons, however, can be very long. For example, axons coming from the spinal cord and traveling to the feet can be 75 to 100 centimeters long! The nerve impulse continues to travel along the afferent neuron until it reaches a nerve cell in the dorsal root ganglion. From here, the nerve impulse travels to a nerve cell in the spinal cord. The nerve impulse continues to travel up the neuron within the spinal cord until it reaches the brain, where it is interpreted and felt as a sensation. Then the brain sends motor neurons back to the receptors, telling them how to respond.

Page 28

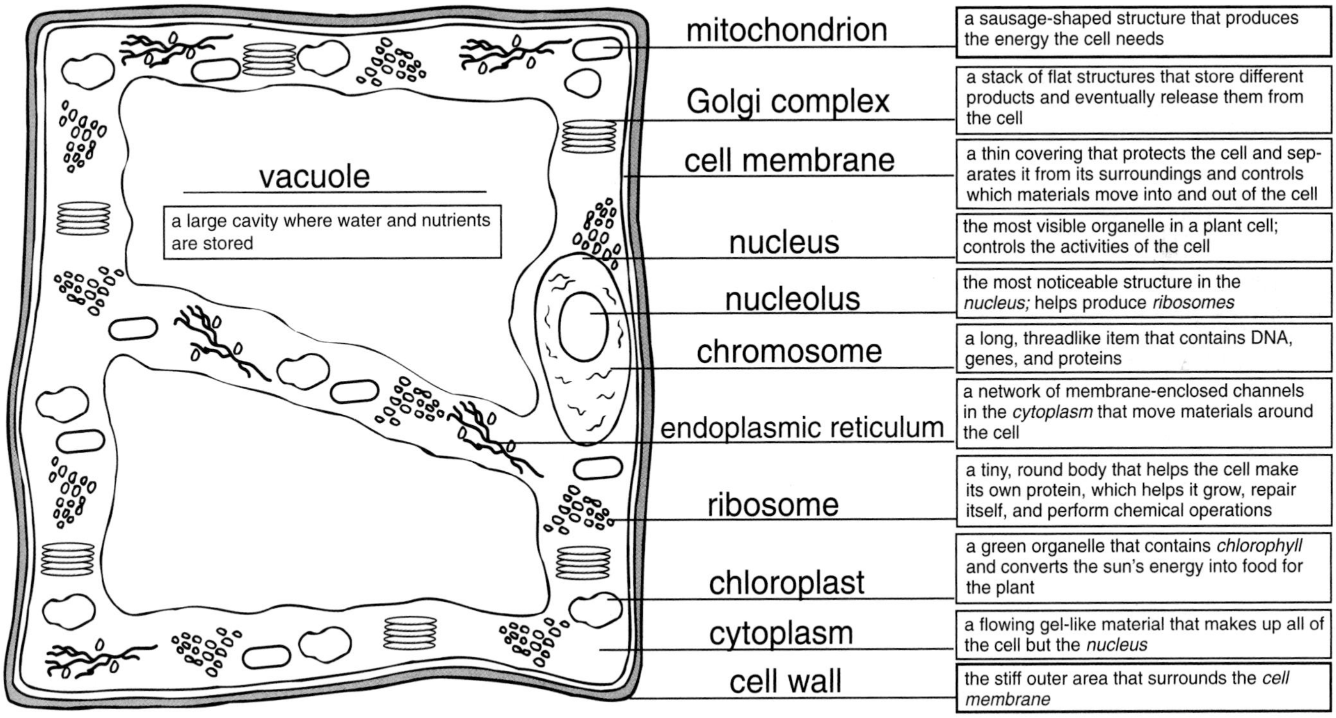

vacuole
a large cavity where water and nutrients are stored

mitochondrion
a sausage-shaped structure that produces the energy the cell needs

Golgi complex
a stack of flat structures that store different products and eventually release them from the cell

cell membrane
a thin covering that protects the cell and separates it from its surroundings and controls which materials move into and out of the cell

nucleus
the most visible organelle in a plant cell; controls the activities of the cell

nucleolus
the most noticeable structure in the *nucleus;* helps produce *ribosomes*

chromosome
a long, threadlike item that contains DNA, genes, and proteins

endoplasmic reticulum
a network of membrane-enclosed channels in the *cytoplasm* that move materials around the cell

ribosome
a tiny, round body that helps the cell make its own protein, which helps it grow, repair itself, and perform chemical operations

chloroplast
a green organelle that contains *chlorophyll* and converts the sun's energy into food for the plant

cytoplasm
a flowing gel-like material that makes up all of the cell but the *nucleus*

cell wall
the stiff outer area that surrounds the *cell membrane*

Page 29

Order may vary.

Flowers: broccoli floret, cauliflower
Roots: carrot, radish, sweet potato
Bulbs: onion, garlic
Leaves: cabbage, lettuce, spinach
Fruits: apple, cucumber, pumpkin, tomatoes, pepper
Stems: asparagus, green onion, celery stalk
Seeds: peas, beans, corn, peanut

Page 30

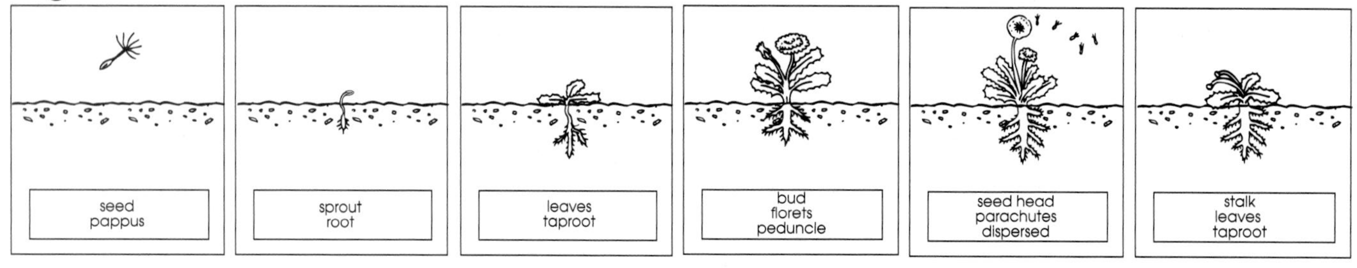

A dandelion seed floats through the air on the wind. The fluffy white circle of hairs is called the *pappus*.

Roots dig into the soil and a dandelion begins to sprout.

A dandelion's root is called a *taproot*. Soon leaves develop and begin to spread.

A bud appears and a cluster of many tiny flowers, called *florets*, bursts from the bud. Each flower cluster is held up by a stalk called a *peduncle*.

The flower cluster closes up and then a seed head opens. The seeds in the head contain parachutes that are dispersed by the wind.

The dandelion flower and stalk die, but the leaves and taproot live on.

Page 32

Part One:

1. sun, transpiration, evaporates
2. stomata
3. absorbs, cool
4. roots, water
5. stem, xylem vessels

Part Two:

1. perspiration
2. pores
3. meat, cheese, and bread
4. arteries and veins
5. vitamins and minerals

Bonus: Analogies will vary.

Page 33

1. poison ivy
2. jack-in-the-pulpit
3. belladonna
4. ragweed
5. oleander
6. rhubarb
7. mushroom
8. dogbane
9. hemlock

NEVER EAT OR CHEW A PLANT YOU DON'T KNOW WELL.

Page 34

1. propagation
2. carbon dioxide
3. seeds
4. nitrogen
5. maturity
6. germination
7. nutrients
8. succession
9. dispersal
10. decompose

Page 35

Students' explanations will vary but should mention that humans get their oxygen from plants during photosynthesis. Students should also mention that plants get the carbon dioxide they need to make food from the air that humans breathe out during respiration.

Bonus: *Anaerobic* means respiration without air. Sentences will vary.

Page 36

1. air
2. soil
3. light
4. light and air
5. water
6. water and air
7. soil
8. light and air
9. soil
10. water

Bonus: Illustrations will vary.

Page 44

Role 1: Ellen Echinodermata
Role 2: Molly Mollusca
Role 3: Arthur Arthropoda
Role 4: Annie Annelida
Role 5: Paul Platyhelminthes
Role 6: Polly Porifera
Role 7: Sid Coelenterate
Role 8: Nan Nematoda

Page 46

Answers will vary but should include the following:

1. Patches of white, green, or black material started to form on the foods. The patches look soft and cottony.
2. Decomposers were in the air inside the jar. Another way decomposers could have gotten into the jar is by water, especially if any of the food items had been washed in water before they were placed in the jar.
3. More decomposition will probably take place during that time.

Page 47

1. musk ox—tundra
2. toucan—tropical forest
3. Arctic hare—tundra
4. ground squirrel—desert
5. opossum—temperate forest
6. kangaroo rat—desert
7. gnu—grassland
8. gibbon—tropical forest
9. lion—grassland
10. salamander—temperate forest

Adaptation examples will vary.

1. thick fur to keep it warm
2. wings to fly from tree to tree
3. short ears and short tail to keep in body heat
4. estivates, or sleeps through the summer, to keep cool
5. small body to move easily through the underbrush
6. spends days in a burrow to keep cool
7. ability to travel long distances to find grass to eat
8. long arms used to swing from tree to tree
9. sharp canine teeth for killing and then tearing prey
10. hides under leaves and rocks to find and eat insects

It thinks it's rather P L A I N!

Page 49

1. D	3. C	5. F	7. H	9. J
2. G	4. E	6. I	8. A	10. B

Bonus: Answers will vary. Accept reasonable responses.

Page 66

"What a Stretch!"

Conclusion: The more the rubber band is stretched, the farther it will travel when released.

Page 82

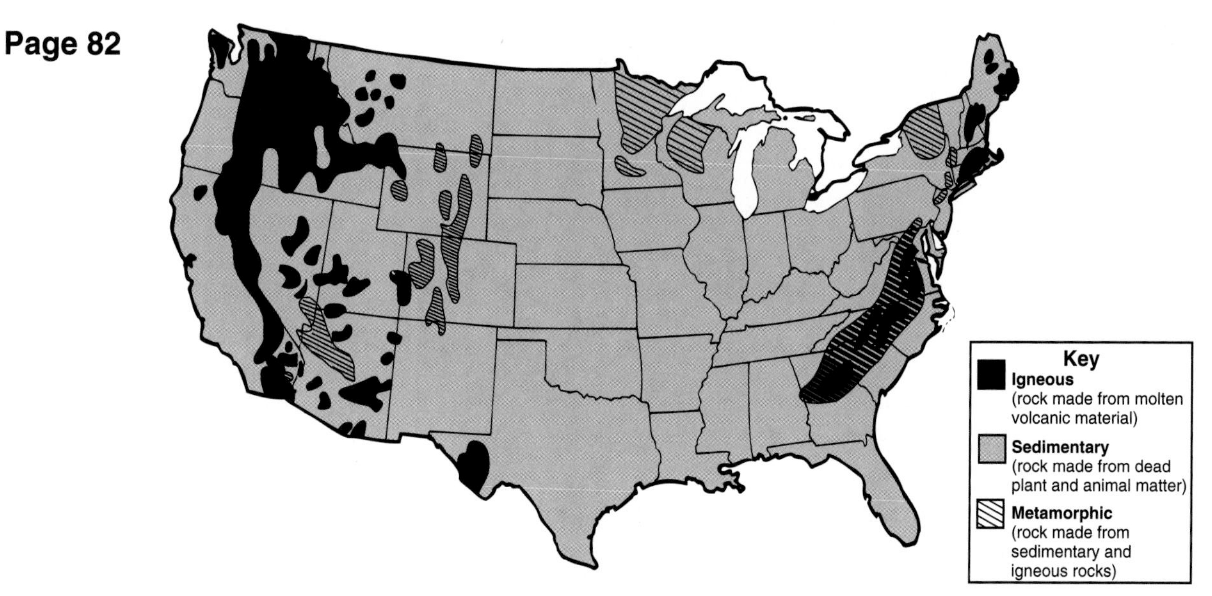

Page 93

Definitions will vary.

1.	Cirrocumulus	curl + heap	Cirrocumulus clouds are like puffy curls of cotton and hold ice crystals.
2.	Cumulonimbus	heap + rain	Cumulonimbus clouds are large, fluffy storm clouds.
3.	Altocumulus	higher + heap	Altocumulus clouds are heaps of clouds seen higher in the sky.
4.	Nimbostratus	rain + spread out	Nimbostratus clouds are thick rain clouds spread across the sky.
5.	Cirrostratus	curl + spread out	Cirrostratus clouds are curls of high clouds spread out across the sky.
6.	Altostratus	higher + spread out	Altostratus clouds are higher, thick clouds spread out across the sky.
7.	Stratocumulus	spread out + heap	Stratocumulus clouds are puffy clouds spread out across the sky.

Page 97

Beaufort Scale 0—It's so calm. It feels like there is no wind at all.
Beaufort Scale 1—The smoke from that chimney is slowly blowing toward the north.
Beaufort Scale 2—The light breeze feels nice and cool.
Beaufort Scale 3—The leaves in that yard are being blown around.
Beaufort Scale 5—That small maple tree is swaying gently.
Beaufort Scale 6—You may need a stronger umbrella; the winds are 29 MPH!
Beaufort Scale 7—That whole tree is moving in the 34-MPH wind.
Beaufort Scale 8—Some of the twigs are breaking off of that tree.
Beaufort Scale 9—That house just lost three shingles from its roof!
Beaufort Scale 10—That tree just pulled right out of the ground!
Beaufort Scale 11—The storm's winds are up to 69 MPH and are causing a lot of damage!
Beaufort Scale 12–17—Evacuate immediately! The winds are over 80 MPH!

Page 99

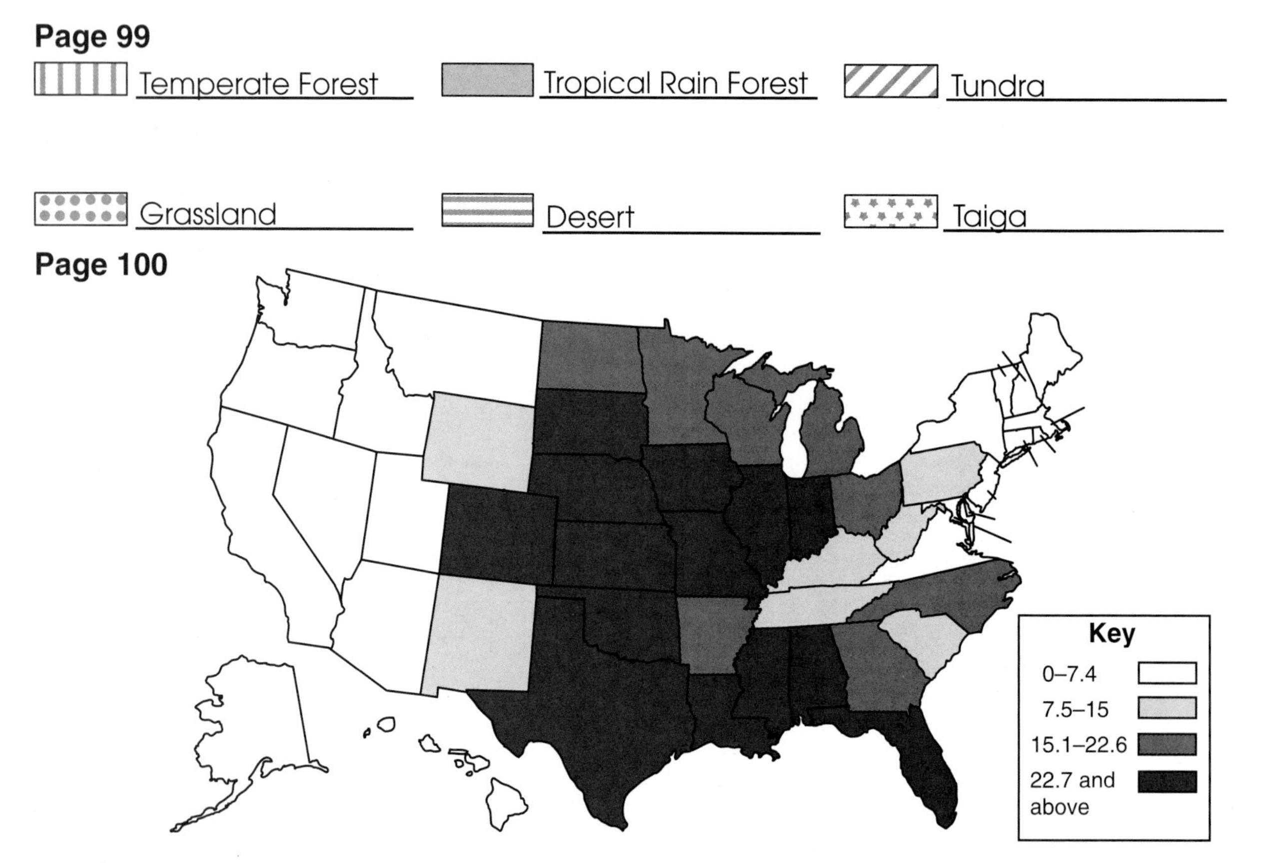

1. Answers will vary but should include those states in the 22.7 and above range that are close to each other in location.
2. Florida
3. Answers will vary but should include that the states in Tornado Alley are plains states. This is good for tornadoes, which move most easily over flat, open spaces.

Page 101

Answers will vary.

1. the Bahamas
2. Puerto Rico, because Windy was much closer to it than to the Dominican Republic
3. the prevailing westerlies
4. Answers will vary but should include that Florida was in some danger, but people would probably not have been evacuated until there was more of a threat.

Page 102

1–4.

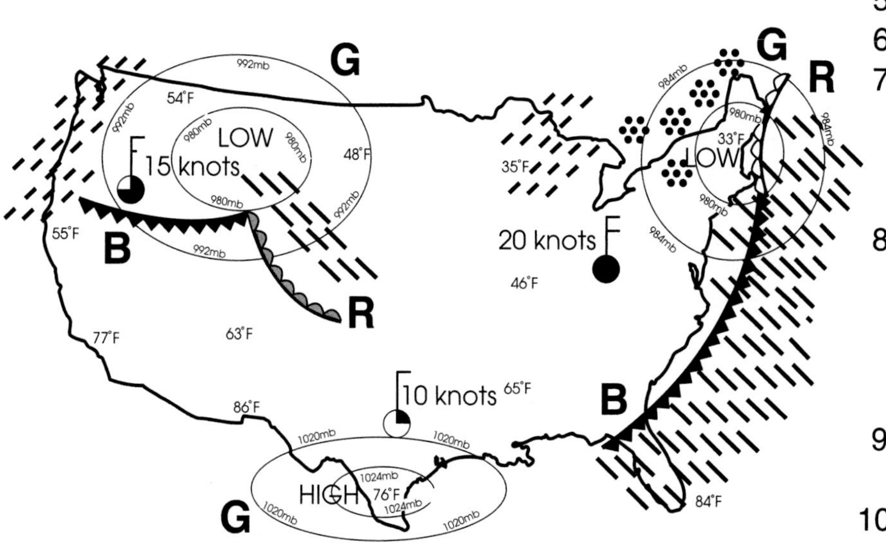

5. the eastern half
6. the East
7. 20 knots; These winds are in the Ohio-Kentucky-West Virginia area, southwest of a low-pressure area.
8. The area north of the Great Plains and all of the east coast are getting rain. The northwest coast and the Great Lakes area are getting showers.
9. The western parts of the Northeast
10. highest temperature, 86°F; lowest temperature, 33°F

Page 111

1 Planet	2 Distance From the Sun (in kilometers)	3 New Scale (10 million kilometers = 1 millimeter)	4 New Scale Rounded to Nearest Whole Number (in millimeters)
Mercury	57,900,000.0	5.7900000	6
Venus	108,200,000.0	10.8200000	11
Earth	149,600,000.0	14.9600000	15
Mars	227,900,000.0	22.7900000	23
Jupiter	778,400,000.0	77.8400000	78
Saturn	1,429,400,000.0	142.9400000	143
Uranus	2,875,000,000.0	287.5000000	288
Neptune	4,504,300,000.0	450.4300000	450

Page 115

I.
1. 9.5 lb.
2. 22.75 lb.
3. 25 lb.
4. 9.5 lb.
5. 63.5 lb.
6. 26.75 lb.
7. 22.5 lb.
8. 28.75 lb.

II. Answers will vary.

Page 116

1. Mercury, Venus, Earth, Mars, Jupiter, Saturn, Uranus, Neptune
2. The sun's gravitational pull keeps each planet in a constant orbit around the sun.
3. Jupiter, Mercury
4. A planet's day is determined by the amount of time it takes that planet to make one rotation.
5. Answers will vary.
6. A planet's year is determined by the amount of time it takes that planet to make one orbit around the sun.
7. Saturn's rings appear different due to the position of Earth and Saturn's orbital position, so the angle at which we view it looks different.
8. Mars is sometimes called the red planet because it shines with a reddish orange color from the iron oxide on its surface.

Page 117

1. 387 million miles
2. 900 million miles
3. Earth to Mars
4. Jupiter to Venus
5. 1,707 million miles more
6. 2,733 million miles
7. Jupiter; 420 million miles closer
8. 100 million miles closer

Page 119

1. lunar
2. satellite
3. a. orbit
   b. eclipse
4. a. air
   b. Moonlight
5. a. maria
   b. lava
6. meteorites
7. 4.6 billion
8. gravity
9. tides
10. astronaut

Page 120

1. Comet Flaugergues, Great Comet, Great September Comet, Comet Ikeya-Seki, Comet Bennett, Comet West
2. 2062 (Astronomers are actually predicting Halley's Comet to return in 2061. The time of the comet's orbit fluctuates, but on average it takes 76 years.)
3. a. 2845 b. 2017 c. 3647 d. 4911 e. 2356
4. Students' tables should be colored as follows:

Yellow	Blue
Halley's Comet	Comet Flaugergues
Comet Swift-Tuttle	Great Comet
Comet Tempel-Tuttle	Great September Comet
Biela's Comet	Comet Ikeya-Seki
Encke's Comet	Comet Bennett
Comet Schwassmann-Wachmann 1	Comet West

Page 121

1. (Asteroids) is, Disagree
2. (Ceres) are, Disagree
3. (asteroids) look, Agree
4. (paths) brings, Disagree
5. (Scientists) study, Agree
6. (Asteroids) vary, Agree
7. (photographs) were taken, Agree
8. (planets and planetoids) is, Disagree
9. (family) are made up, Disagree
10. (Ceres) makes up, Agree
11. (People) has traveled, Disagree
12. (asteroids) are found, Agree
13. (Astronomers) uses, Disagree

ASTEROIDS ASTOUND ASTRONOMERS!